T0135712

Model reduction and optimal control in field-flow fractionation

Dissertation

zur Erlangung des akademischen Grades

doctor rerum naturalium
(Dr. rer. nat.)

eingereicht an der

Mathematisch-Naturwissenschaftlich-Technischen Fakultät

der Universität Augsburg

von

Carina Willbold

Augsburg, Juli 2016

Universität
Augsburg
University

Augsburger Schriften zur Mathematik, Physik und Informatik
Band 34

Edited by:
Professor Dr. B. Schmidt
Professor Dr. B. Aulbach
Professor Dr. F. Pukelsheim
Professor Dr. W. Reif
Professor Dr. D. Vollhardt

Bibliographic information published by the Deutsche Nationalbibliothek

The Deutsche Nationalbibliothek lists this publication in the Deutsche Nationalbibliografie; detailed bibliographic data are available in the Internet at http://dnb.d-nb.de .

ISBN 978-3-8325-4506-2
ISSN 1611-4256

Logos Verlag Berlin GmbH
Comeniushof, Gubener Str. 47,
10243 Berlin
Tel.: +49 030 42 85 10 90
Fax: +49 030 42 85 10 92
INTERNET: http://www.logos-verlag.de

Gutachter: 1. Prof. Dr. Tatjana Stykel
2. Prof. Dr. Malte Peter

Tag der mündlichen Prüfung: 30.09.2016

Acknowledgements

I want to thank my supervisor Prof. Dr. Tatjana Stykel for the chance of working with the fascinating topics I did my research in. I also want to thank her for her patience and that she always had an open ear for all questions.

Furthermore, I'm very grateful to Prof. Dr. Marc Nieper-Wißkirchen for providing me inspiration for a proof of a specific result. In addition, I thank my colleague Johanna Kerler-Back for answering questions and giving me advice.

I want to thank Ingo Blechschmidt and Dr. Peter Uebele for their substantial support, for their counterexamples and for the discussion of mathematical problems at all times. In addition, Felix Geißler, Dr. Hedwig Heizinger, Kathrin Helmsauer, Lisa Reischmann, and Caren Schinko receive my deepest gratitude for their special encouragement.

Contents

1 Introduction

We consider optimization problems governed by a coupled system of equations in the context of field flow fractionation, a technique used for the separation of particles of different sizes in a microfluidic flow. In order to solve the occurring optimization problems efficiently, model reduction techniques for the accompanying equations will be developed and used.

In Chapter 2, we describe the special case of asymmetric flow-field-flow fractionation (see [68]) and the different phases of this process. Definitions, notations and results that are used in this work can be found in Chaper 3. Then, in Chapter 4, we formulate the problem as an optimization problem and we discuss the occuring partial differential equations, namely the Stokes–Brinkman and advection-diffusion equations. Furthermore, we derive in Chapter 5 a spatial discretization of the problem using finite element methods.

Chapter 6 deals with the existence of minimizers for the optimal control problem and the differentiability of the objective functional. We prove the existence of an optimal control for the semidiscretized case. The difficulty lies in the choice of the function spaces, as they have to be fitted for the computation of the adjoint equations. We show furthermore as a preparation for the use of the Lagrange method that the solution operators are sufficiently differentiable. In addition, we obtain an a posteriori error estimate.

Afterwards, in Chapter 7, the existence of Lagrange multipliers is proved and we use the Lagrange framework to compute the adjoint equations and the gradient for the optimization problem. Showing the surjectivity of the differential of the constraints directly fails due to the regularity requirements for the right-hand side of the Stokes–Brinkman equation. In order to guarantee existence and sufficient regularity of Lagrange multipliers, we consider an auxiliary problem which leads to necessary optimality conditions of first order. The basic idea lies in the transformation of a differential-algebraic equation into an ordinary differential equation, where the output behaviour of the system remains unchanged under this transformation. We can show that the Lagrange multipliers and the necessary optimality conditions of the original problem emerge from this approach.

In order to deal with the computational effort, Chapter 8 covers model reduction techniques for the semi-discretized equations. We will focus on techniques for linear equations, recall the Iterative Rational Krylov algorithm and present an extension of balanced truncation with many outputs to the Stokes–Brinkman equation. For the advection-diffusion equation, we recall Proper Orthogonal Decomposition and Discrete Enpirical Interpolation (DEIM). We furthermore reformulate DEIM for matrices in an efficient way. Finally, in Chapter 9, numerical results will be presented.

2 Description of the asymmetric field-flow fractionation

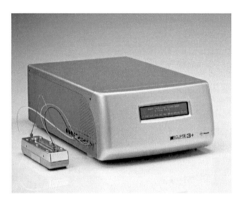

Figure 2.1: The system *Eclipse* developed by Wyatt Germany GmbH, see [25].

We start with a description of field-flow fractionation. Field-flow fractionation (FFF) is a family of techniques for the separation of particles and macromolecules in microfluidic flows. In this work, we consider the special case of asymmetric flow field-flow fractionation (AF^4), the most used variant of the FFF techniques.

The separation process is performed by the analyser Eclipse3, see Figure 2.1, which was developed and is sold by the Wyatt Germany GmbH. The AF^4 takes place in the left component in Figure 2.1 which contains a seperation channel and a membrane. The separation of the particles takes place in a thin channel Ω_1 with a permeable membrane Ω_2 made of a porous frit material as shown in Figures 2.2 and 2.3. The separation process itself includes three steps: injection, focussing and elution. At the first step, the liquid is injected through the injection tube at the top of the channel. There is a crossflow through the membrane and outflow at the bottom boundary. When the flow is balanced, the analyte is injected. The goal of the focussing phase is to concentrate the analyte in a thin band and move it in a carrier fluid towards a given position.

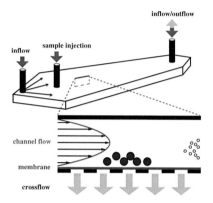

Figure 2.2: The focussing phase of the AF^4.

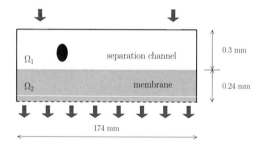

Figure 2.3: The separation channel and the membrane.

Finally, after the focussing phase the separation of the particles occurs then in the elution phase, when a parabolic flow profile is created within the channel. The smaller partices are transported much more rapidly along the channel and eluted earlier than the larger ones. Figure 2.3 shows the geometry of the channel and the membrane included in the Ecplise system.

In this work, only the focussing phase will be considered. For a more detailed view on the whole process of AF^4, we refer to [68].

3 Preliminaries

In this short chapter, we recall some definitions and notions used in this work.

For a matrix $A \in \mathbb{R}^{n \times m}$, we denote the corresponding linear mapping by $L_A : \mathbb{R}^m \to \mathbb{R}^n$, where $L_A(x) = Ax$ for $x \in \mathbb{R}^m$.

Now we introduce some notations of Banach and Hilbert spaces. Let $\Omega \subset \mathbb{R}^d$ be some given domain. The Banach space $(C^p(\Omega))^q$ with $p \in \mathbb{N}_0, q \in \mathbb{N}$ denotes the space of vector-valued functions which are p times continuously differentiable, i.e.

$$(C^p(\Omega))^q = \{f : \Omega \to \mathbb{R}^q \mid f \text{ is } p \text{ times continuously differentiable}\}.$$

In the case $p \in \mathbb{N}_0$, the norm is given by

$$\|f\|_{(C^p(\Omega))^q} = \sum_{i=0}^{p} \sup_{x \in \Omega} \|f^{(i)}(x)\|_2. \tag{3.1}$$

In case of $p = \infty$ we define

$$(C^\infty(\Omega))^q = \{f : \Omega \to \mathbb{R}^q \mid f \text{ is arbitrarily often continuously differentiable}\},$$

which is not a Banach space, but a Fréchet space, see [20].

Let now $\partial\Omega$ denote the boundary of the domain Ω. Then the space $(C_0^p(\Omega))^q$ with $p \in \mathbb{N}_0 \cup \infty$ and $q \in \mathbb{N}$ is defined as the space of p times continuously differentiable functions with local support. In the case $p < \infty$, this space is a Banach space with the norm (3.1), where as in case $p = \infty$ it still remains a Fréchet space.

The space $(L^p(\Omega))^q$ for $p \in \mathbb{N} \cup \{\infty\}$ and $q \in \mathbb{N}$ denotes the space of p-integrable functions on Ω,

$$(L^p(\Omega))^q := \{v : \Omega \to \mathbb{R}^q \mid v \text{ is measurable and } \int_\Omega \|v(x)\|_2^p \, dx < \infty\}.$$

This space is a Banach space with respect to the norm

$$\|f\|_{(L^p(\Omega))^q} = \left(\int_\Omega \|f(x)\|_2^p \, dx \right)^{\frac{1}{p}}$$

for $p < \infty$ and

$$\|f\|_{(L^\infty(\Omega))^q} = \operatorname{ess\,sup}_{x \in \Omega} \|f(x)\|_2$$

for $p = \infty$ using the essential supremum.

In case $p = 2$, $(L^2(\Omega))^q$ is with respect to the scalar product

$$(f, g)_{(L^2(\Omega))^q} = \int_\Omega (f(x), g(x))_2 \, \mathrm{d}x$$

a Hilbert space.

Furthermore, the space $(L^{1,\mathrm{loc}}(\Omega))^q$ for $q \in \mathbb{N}$ denotes the locally integrable functions on Ω, i.e.

$$(L^{1,\mathrm{loc}}(\Omega))^q = \{f : \Omega \to \mathbb{R}^q \text{ measurable} \mid f|_K \in L_1(K) \text{ for all } K \subset \Omega, K \text{ compact}\}.$$

We briefly recall the definition of a weak derivative. Let $u \in (L^{1,\mathrm{loc}}(\Omega))^q$.
We call $v \in (L^{1,\mathrm{loc}}(\Omega))^q$ the α-th weak derivative of u for the multi-index $\alpha \in \mathbb{N}_0^d$ if

$$\int_\Omega (u(x), D^\alpha \varphi(x))_2 \, \mathrm{d}x = (-1)^{|\alpha|} \int_\Omega (v(x), \varphi(x))_2 \, \mathrm{d}x$$

for all $\varphi \in (C_0^\infty(\Omega))^q$, where $D^\alpha f = \frac{\partial^{|\alpha|} f}{\partial x_1^{\alpha_1} \ldots \partial x_d^{\alpha_d}}$.

The Hilbert space $(H^p(\Omega))^q$ for $p, q \in \mathbb{N}$ is defined as

$$(H^p(\Omega))^q = \{v : \Omega \to \mathbb{R}^q \mid v \text{ is } p \text{ times weakly differentiable with } D^\alpha v \in (L^2(\Omega))^q \text{ for } |\alpha| \leq p\}$$

with respect to the scalar product and norm

$$(f, g)_{(H^p(\Omega))^q} = \sum_{|\alpha| \leq p} (D^\alpha f, D^\alpha g)_{(L^2(\Omega))^q},$$

$$\|f\|_{(H^p(\Omega))^q} = \left(\sum_{|\alpha| \leq p} \|D^\alpha f\|_2^2 \right)^{\frac{1}{2}},$$

where $D^\alpha f, D^\alpha g$ denote the α-th weak derivative of f, g, respectively.

Let now H denote a Hilbert space with a norm $\|\cdot\|_H$ and let $(0, T) \subset \mathbb{R}$. Then the Hilbert spaces $L^2(0, T; H)$ and $H^p(0, T; H)$ for $p \in \mathbb{N}$ are defined as

$$L^2(0, T; H) = \{v : (0, T) \to H \mid \int_0^T \|v(t)\|_H^2 \, \mathrm{d}t < \infty\},$$
$$H^p(0, T; H) = \{v : (0, T) \to H \mid D^\alpha v \in L^2(0, T; H) \text{ for } |\alpha| \leq p\}.$$

Furthermore, the space $C^p(0, T; H)$ for $p \in \mathbb{N}$ is defined as

$$C^p(0, T; H) = \{v : (0, T) \to H \mid v \text{ is } p \text{ times continuously differentiable}\}.$$

Let V denote a Banach space. Then the dual space V^* is defined as

$$V^* = \{a : V \to \mathbb{R} \mid a \text{ is linear and continuous}\}.$$

As is common, the (topological) dual space of $H_0^1(\Omega)$ is denoted by $H^{-1}(\Omega)$ in this work.

Let H be a separable Hilbert space, V a Banach space which lies dense in H. We identify H, H^* by the canonical isomorphism. Then a Gelfand triple is of the form

$$V \overset{J}{\hookrightarrow} H \cong H^* \overset{J^*}{\hookrightarrow} V^*.$$

Here V^*, H^* denote the dual spaces of V, H respectively. The operator $J : V \hookrightarrow H$ is an injective bounded operator, and, because $J(V)$ is dense in H, the dual operator $J^* : H^* \hookrightarrow V^*$ is also injective.

We recall now several notations of special derivatives. Let therefore $\mathbf{v}, \boldsymbol{\varphi} : \Omega \to \mathbb{R}^d$ be weakly differentiable functions. Then we define

$$(\nabla \mathbf{v})_{ij} := \frac{\partial \mathbf{v}_j}{\partial x_i}$$

$$\nabla \cdot \mathbf{v} := \sum_{i=1}^{d} \frac{\partial \mathbf{v}_i}{\partial x_i}$$

$$\Delta \mathbf{v} := \nabla \cdot (\nabla \mathbf{v})$$

$$\nabla \mathbf{v} : \nabla \boldsymbol{\varphi} := \sum_{i=1}^{d} \sum_{j=1}^{d} \frac{\partial \mathbf{v}_i}{\partial x_j} \frac{\partial \boldsymbol{\varphi}_i}{\partial x_j}$$

The boundary derivative $\frac{\partial \mathbf{v}}{\partial \mathbf{n}_{\partial \Omega}}$ is defined as

$$\frac{\partial \mathbf{v}}{\partial \mathbf{n}_{\partial \Omega}} := (\nabla \mathbf{v}) \mathbf{n}_{\partial \Omega}$$

where $\mathbf{n}_{\partial \Omega}$ denotes the outer normal vector on the boundary $\partial \Omega$ of Ω.

Now we present some definitions and results which will be used later on.

Definition 3.1. A function $\mathbf{w} \in L^2(0, T; V^*)$ is a *weak derivative* of $\mathbf{u} \in L^2(0, T; V)$ if

$$\int_0^T \langle \mathbf{w}(t), \mathbf{s} \rangle \phi(t) \, \mathrm{d}t = - \int_0^T (\mathbf{u}(t), \mathbf{s})_H \frac{\mathrm{d}}{\mathrm{d}t} \phi(t) \, \mathrm{d}t$$

holds true for all $\phi \in C_0^\infty(0, T)$ and $\mathbf{s} \in V$. Here $\langle \cdot, \cdot \rangle : V^* \times V \to \mathbb{R}$ denotes the dual pairing, and $(\cdot, \cdot)_H$ is a scalar product in a Hilbert space H.

Definition 3.2. *Let $F : \mathbb{R}^q \to \mathbb{R}^{n \times n}$ be differentiable such that $DF : \mathbb{R}^q \to \mathbb{R}^{n \times n}$ is uniformly continuous. The function F is called uniformly differentiable if for any $\epsilon > 0$, there is a $\delta > 0$ such that*

$$\frac{\|F(x) - F(a) - DF(x)(x - a)\|}{\|x - a\|} < \epsilon$$

holds for all $a, x \in \mathbb{R}^q$ with $\|x - a\| < \delta$.

At last, we recall some useful results which are used in this work.

Lemma 3.3. *(Gronwall's Lemma [32])*

Let $I = [a, b] \subset \mathbb{R}$ and $u : I \to \mathbb{R}$. If there exists a continuous monotonic increasing function $\alpha : I \to \mathbb{R}$ and a continuous function $\beta : I \to [0, \infty)$ such that

$$u(t) \leq \alpha(t) + \int_a^t \beta(s) u(s) \, ds,$$

then

$$u(t) \leq \alpha(t) e^{\int_a^t \beta(s) \, ds}.$$

Theorem 3.4. *(Closed Graph Theorem [62])*

Let X, Y be Banach spaces, and $T : X \to Y$ a linear mapping. Then T is continuous if and only if its graph $G = \{(x, y) \in X \times Y \mid Tx = y\}$ is closed in $X \times Y$.

The next lemma was established in [2].

Lemma 3.5. *(Estimate for ODE solution)*

Let $M \in \mathbb{R}^{n \times n}$ be symmetric and positive definite, $A \in \mathbb{R}^{n \times n}, \mathbf{y}_0 \in \mathbb{R}^n$ and $B \in \mathbb{R}^{n \times m}$. If there exists a constant $\alpha > 0$ such that $\mathbf{v}^T A \mathbf{v} \geq \alpha \mathbf{v}^T M \mathbf{v}$ for all $\mathbf{v} \in \mathbb{R}^n$, then the unique solution of

$$M \frac{d}{dt} \mathbf{y}(t) + A \mathbf{y}(t) = B \mathbf{u}(t),$$

$$\mathbf{y}(0) = \mathbf{y}_0$$

obeys

$$\|\mathbf{y}\|_{L^2} \leq \frac{\sqrt{2}}{\sqrt{\alpha}} \|M^{-\frac{1}{2}}\| \|M^{\frac{1}{2}}\| \|\mathbf{y}_0\| + \frac{2}{\alpha} \|M^{-1}\| \|B\mathbf{u}\|_{L^2}.$$

4 The optimal control problem

In this chapter, we present an optimal control problem arising in the focussing phase of the AF[4]. We use the Stokes–Brinkman equation to model the fluid flow in the channel and the advection-diffusion equation to describe the concentration of the analytes. The goal of the optimal control problem is to find an inflow profile such that the distribution of the particles approximates a desired concentration profile at a fixed time.

4.1 The Stokes–Brinkman equation

Let $\Omega = \Omega_1 \cup \Omega_2 \subset \mathbb{R}^2$ be a rectangular domain consisting of a channel Ω_1 and a membrane Ω_2 as shown in Figure 4.1. The flow of the fluid in the domain Ω is described by the incompressible Stokes–Brinkman equation

$$\rho\frac{\partial \mathbf{v}}{\partial t} - \nu\Delta\mathbf{v} + \nu\chi_{\Omega_2}K^{-1}\mathbf{v} + \nabla p = 0 \qquad \text{in} \quad \Omega \times (0,T), \tag{4.1}$$

$$\nabla \cdot \mathbf{v} = 0 \qquad \text{in} \quad \Omega \times (0,T), \tag{4.2}$$

$$\mathbf{v} = \mathbf{v}_{\text{in}}^{(i)} \qquad \text{on} \quad \Gamma_{\text{in}}^{(i)} \times (0,T), i = 1,2, \tag{4.3}$$

$$\mathbf{v} = 0 \qquad \text{on} \quad \Gamma_{\text{lat}} \times (0,T), \tag{4.4}$$

$$\nu\frac{\partial \mathbf{v}}{\partial \mathbf{n}_{\Gamma_{\text{out}}}} - p\mathbf{n}_{\Gamma_{\text{out}}} = 0 \qquad \text{on} \quad \Gamma_{\text{out}} \times (0,T), \tag{4.5}$$

$$\mathbf{v}(\cdot,0) = \mathbf{v}_0 \qquad \text{in} \quad \Omega, \tag{4.6}$$

where $\mathbf{v}(x,t) \in \mathbb{R}^2$ denotes the velocity, $p(x,t) \in \mathbb{R}$ denotes the pressure, ρ and ν are the density and viscosity constants, respectively, and K is the permeability of the membrane. Figure 4.1 shows where the boundary conditions are posed in detail. On the top of the channel, there are two inflow boundaries $\Gamma_{\text{in}}^{(1)}$ and $\Gamma_{\text{in}}^{(2)}$, where the inflow conditions (4.3) are stated, the outflow at the bottom Γ_{out} of the membrane is described by the do-nothing outflow boundary condition (4.5), see [61]. On the remaining boundary Γ_{lat} we use the no-slip boundary condition (4.4), see [23].

The Brinkman term $\nu\chi_{\Omega_2}K^{-1}\mathbf{v}$ is nontrivial only on the membrane Ω_2 due to the characteristic function $\chi_{\Omega_2} : \Omega \to \mathbb{R}$ given by

$$\chi_{\Omega_2}(x) = \begin{cases} 1, & x \in \Omega_2, \\ 0, & x \notin \Omega_2 \end{cases}$$

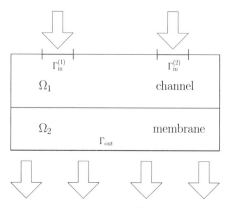

Figure 4.1: The flow domain $\Omega = \Omega_1 \cup \Omega_2 \subset \mathbb{R}^2$ for the Stokes–Brinkman equation.

The membrane is treated as semi-permeable, meaning that only the fluid can cross the membrane, but the particles not.

For the description of the two inflow profiles $\mathbf{v}_{\text{in}}^{(i)} = [v_1^{(i)}, v_2^{(i)}]^T, i = 1, 2$, we use the parabolic functions

$$v_1^{(i)}(x, t) = 0, \tag{4.7}$$

$$v_2^{(i)}(x, t) = -u_i(t)(x_1 - b_1)(b_2 - x_1) \tag{4.8}$$

for $t \in [0, T]$ and $x = [x_1, x_2]^T \in \Gamma_{\text{in}}^{(i)}, i = 1, 2$. The function $\mathbf{u}(t) = [u_1(t), u_2(t)]^T$ is considered as the control.

For a derivation of the Stokes–Brinkman equation, we refer to [15]. The Stokes–Brinkman equation as constraint in an optimal control problem has already been studied, for example in [12].

The weak formulation

In order to use the finite element method (FEM), e.g. [33], for a spatial discretization we have to compute the weak formulation of the Stokes–Brinkman equation (4.1)–(4.6) at first. For this purpose we consider the Sobolev spaces

$$V = \{\mathbf{w} \in H^1(\Omega)^2 \,|\, \mathbf{w}|_{\Gamma_{\text{lat}} \cup \Gamma_{\text{in}}} = 0\},$$

$$U = \{\mathbf{w} \in H^1(\Omega)^2 \,|\, \mathbf{w}|_{\Gamma_{\text{lat}}} = 0, \, \mathbf{w}|_{\Gamma_{\text{in}}^i} = \mathbf{v}_{\text{in}}^{(i)}, i = 1, 2\},$$

$$V(0, T) = H^1(0, T; V^*) \cap L^2(0, T; U),$$

$$P(0, T) = L^2(0, T; L^2(\Omega)),$$

where $\Gamma_{\text{in}} = \Gamma_{\text{in}}^{(1)} \cup \Gamma_{\text{in}}^{(2)}$.

The differentiability condition in $V(0, T)$ should be understood in a weak way in the sense of the Gelfand triple [30]

$$H^1_{0, \Gamma_{\text{lat}} \cup \Gamma_{\text{in}}}(\Omega)^2 \hookrightarrow L^2(\Omega)^2 \cong (L^2(\Omega)^2)^* \hookrightarrow H^{-1}_{0, \Gamma_{\text{lat}} \cup \Gamma_{\text{in}}}(\Omega)^2.$$

Assume that $\mathbf{u} \in H^3(0, T; \mathbb{R}^2)$ and $\mathbf{v}_0 \in L^2(\Omega)$. We want to find a solution $(\mathbf{v}, p)$ of the Stokes–Brinkman equation (4.1)–(4.6) in $V(0, T) \times P(0, T)$. In order to establish the weak formulation of the Stokes–Brinkman equation, we multiply the equations (4.1) and (4.2) with test functions $\boldsymbol{\varphi} \in V$ and $\psi \in L^2(\Omega)$, respectively, and integrate them over the domain Ω. This leads to

$$\rho \frac{\partial}{\partial t} \int_\Omega \mathbf{v} \cdot \boldsymbol{\varphi} \, dx - \nu \int_\Omega \Delta \mathbf{v} \cdot \boldsymbol{\varphi} \, dx + \nu K^{-1} \int_\Omega \chi_{\Omega_2} \mathbf{v} \cdot \boldsymbol{\varphi} \, dx + \int_\Omega \nabla p \cdot \boldsymbol{\varphi} = 0,$$

$$\int_\Omega (\nabla \cdot \mathbf{v}) \psi \, dx = 0.$$

Integrating by parts we obtain the equations

$$\nu \int_\Omega \Delta \mathbf{v} \cdot \boldsymbol{\varphi} \, dx = -\nu \int_\Omega \nabla \mathbf{v} : \nabla \boldsymbol{\varphi} \, dx + \nu \int_{\Gamma_{\text{out}}} \nabla \mathbf{v} \mathbf{n}_{\Gamma_{\text{out}}} \cdot \boldsymbol{\varphi} \, ds,$$

$$\int_\Omega \nabla p \cdot \boldsymbol{\varphi} \, dx = - \int_\Omega p(\nabla \cdot \boldsymbol{\varphi}) \, dx + \int_{\Gamma_{\text{out}}} p \mathbf{n}_{\Gamma_{out}} \cdot \boldsymbol{\varphi} \, ds.$$

Then using the boundary condition (4.5) we get

$$\rho \frac{\partial}{\partial t} \int_\Omega \mathbf{v} \cdot \boldsymbol{\varphi} \, dx + \nu \int_\Omega \nabla \mathbf{v} : \nabla \boldsymbol{\varphi} \, dx + \nu K^{-1} \int_{\Omega_2} \mathbf{v} \cdot \boldsymbol{\varphi} \, dx - \int_\Omega p(\nabla \cdot \boldsymbol{\varphi}) \, dx = 0, \qquad (4.9)$$

$$\int_\Omega (\nabla \cdot \mathbf{v}) \psi \, dx = 0. \qquad (4.10)$$

Thus we arrive at the following variational problem: Find $(\mathbf{v}, p) \in V(0, T) \times P(0, T)$ such that

$$\frac{\mathrm{d}}{\mathrm{d}t} m_{sb}(\mathbf{v}, \boldsymbol{\varphi}) + a_{sb}(\mathbf{v}, \boldsymbol{\varphi}) - b_{sb}(\boldsymbol{\varphi}, p) = 0,$$

$$b_{sb}(\mathbf{v}, \psi) = 0,$$

$$(\mathbf{v}(\cdot, 0), \boldsymbol{\varphi})_{L^2(\Omega)} = (\mathbf{v}_0, \boldsymbol{\varphi})_{L^2(\Omega)}$$

holds true for all $\boldsymbol{\varphi} \in V$ and $\psi \in L^2(\Omega)$. Here the bilinear forms are given by

$$m_{sb}(\boldsymbol{\phi}, \boldsymbol{\xi}) = \rho \int_\Omega \boldsymbol{\phi} \cdot \boldsymbol{\xi} \, dx,$$

$$a_{sb}(\boldsymbol{\phi}, \boldsymbol{\xi}) = \nu \int_\Omega \nabla \boldsymbol{\phi} : \nabla \boldsymbol{\xi} \, dx + \nu K^{-1} \int_{\Omega_2} \boldsymbol{\phi} \cdot \boldsymbol{\xi} \, dx,$$

$$b_{sb}(\boldsymbol{\phi}, \psi) = \int_\Omega \psi(\nabla \cdot \boldsymbol{\phi}) \, dx.$$

4.2 The advection-diffusion equation

The concentration of the particles in the channel Ω_1 is described by the advection-diffusion equation. For the derivation of the advection-diffusion equation, we refer to [6]. Having the advection-diffusion equation as a constraint in an optimization problem was, for example, treated in [7]. The advection-diffusion equation reads

$$\frac{\partial c}{\partial t} - \nabla \cdot D\nabla c + \mathbf{v} \cdot \nabla c = 0 \qquad \text{in} \quad \Omega_1 \times (0, T), \tag{4.11}$$

$$-D\nabla c \cdot \mathbf{n}_{\Gamma_1} + c\mathbf{v} \cdot \mathbf{n}_{\Gamma_1} = 0 \qquad \text{on} \quad \Gamma_1 \times (0, T), \tag{4.12}$$

$$c(\cdot, 0) = c_0 \qquad \text{in} \quad \Omega_1, \tag{4.13}$$

where $c(x, t) \in \mathbb{R}$ is the concentration of the particles, D is the diffusion coefficient and $\mathbf{v}$ is the velocity vector satisfying the Stokes–Brinkman equation (4.1)–(4.6). Furthermore, c_0 is the intial concentration profile and $\mathbf{n}_{\Gamma_1}$ is the outer normal vector on the boundary Γ_1 of Ω_1. The Neumann boundary condition (4.12) ensures that the particles do not leave the channel. Note that the advection–diffusion equation (4.11)–(4.13) is only posed on Ω_1, because the particles are not allowed to pass the membrane, see Figure 4.1.

The advection–diffusion equation (4.11)–(4.13) takes the velocity $\mathbf{v}$ as an input. This means that we have a coupled system of the Stokes–Brinkman and advection–diffusion equations with the coupling in one direction.

The weak formulation

For the advection–diffusion equation (4.11)–(4.13), we consider the Sobolev space

$$V_2(0, T) = H^1(0, T; H^{-1}(\Omega_1)) \cap L^2(0, T; H^1(\Omega_1)).$$

We want to find a solution c of (4.11)–(4.13) in $V_2(0, T)$ assuming that the initial condition c_0 is a $L^2(\Omega_1)$-function.

Again, we need to establish the weak formulation of (4.11)–(4.13). Multiplying (4.11) with a test function $\varphi \in H^1(\Omega_1)$, and integrating over the domain Ω_1, we obtain

$$\frac{\partial}{\partial t} \int_{\Omega_1} c\varphi \, \mathrm{d}x - \int_{\Omega_1} \nabla \cdot D\nabla c\,\varphi \, \mathrm{d}x + \int_{\Omega_1} \mathbf{v} \cdot \nabla c\,\varphi \, \mathrm{d}x = 0.$$

Integration by parts yields

$$\int_{\Omega_1} \nabla \cdot D\nabla c\,\varphi \, \mathrm{d}x = -D \int_{\Omega_1} \nabla c \cdot \nabla \varphi \, \mathrm{d}x - D \int_{\Gamma_1} (\nabla c \cdot \mathbf{n}_{\Gamma_1})\,\varphi \, \mathrm{d}s,$$

$$\int_{\Omega_1} (\mathbf{v} \cdot \nabla c)\varphi \, \mathrm{d}x = -\int_{\Omega_1} \nabla \varphi \cdot \mathbf{v}c \, \mathrm{d}x + \int_{\Gamma_1} c\varphi(\mathbf{v} \cdot \mathbf{n}_{\Gamma_1}) \, \mathrm{d}s.$$

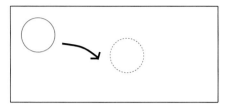

Figure 4.2: The initial concentration profile (blue) and a desired concentration profile (red).

Then taking into account the boundary condition (4.12) we have

$$-D \int_{\Gamma_1} \nabla c \cdot \mathbf{n}_{\Gamma_1} \, \varphi \, \mathrm{d}s + \int_{\Gamma_1} c \varphi \mathbf{v} \cdot \mathbf{n}_{\Gamma_1} \, \mathrm{d}s = 0.$$

The weak formulation reads then: Find $c \in V_2(0, T)$ such that

$$\frac{\mathrm{d}}{\mathrm{d}t} m_{ad}(c, \varphi) + a_{ad}(c, \varphi) = 0,$$
$$(c(\cdot, 0), \varphi)_{L^2(\Omega_1)} = (c_0, \varphi)_{L^2(\Omega_1)}$$

holds true for all $\varphi \in H^1(\Omega_1)$, where the bilinear forms are given by

$$m_{ad}(\xi, \varphi) := \int_{\Omega_1} \xi \, \varphi \, \mathrm{d}x,$$
$$a_{ad}(\xi, \varphi) := D \int_{\Omega_1} \nabla \xi \cdot \nabla \varphi \, \mathrm{d}x - \int_{\Omega_1} \nabla \varphi \cdot \mathbf{v} \xi \, \mathrm{d}x.$$

4.3 The optimization problem

The goal of the optimization problem is to find an inflow velocity $\mathbf{u} \in H^3(0, T; \mathbb{R}^2)$ such that the distance between the concentration of the analyte at the final time T and a desired concentration profile $c^{\mathrm{foc}} \in C(\Omega_1, \mathbb{R})$, see Figure 4.2, is minimized up to a regularisation term.

The optimization problem reads

$$\text{Minimize } J(\mathbf{u}) := \frac{1}{2} \| c(T) - c^{\mathrm{foc}} \|_{L^2(\Omega_1)}^2 + \frac{\sigma}{2} \| \mathbf{u} \|_{H^3(0, T; \mathbb{R}^2)}^2 \tag{4.14}$$

such that the equations

$$\frac{\mathrm{d}}{\mathrm{d}t} m_{sb}(\mathbf{v}, \boldsymbol{\varphi}) + a_{sb}(\mathbf{v}, \boldsymbol{\varphi}) - b_{sb}(\boldsymbol{\varphi}, p) = 0, \tag{4.15}$$
$$b_{sb}(\mathbf{v}, \psi) = 0, \tag{4.16}$$
$$(\mathbf{v}(\cdot, 0), \boldsymbol{\varphi})_{L^2(\Omega)} = (\mathbf{v}_0, \boldsymbol{\varphi})_{L^2(\Omega)}, \tag{4.17}$$

19

and

$$\frac{\mathrm{d}}{\mathrm{d}t}m_{ad}(c,\varphi) + a_{ad}(c,\varphi) = 0, \tag{4.18}$$

$$(c(\cdot,0),\varphi)_{L^2(\Omega_1)} = (c_0,\varphi)_{L^2(\Omega_1)} \tag{4.19}$$

hold true for all $\varphi \in V, \psi \in L^2(\Omega)$ and for all $\varphi \in H^1(\Omega_1)$.

In this chapter, we have presented the optimal control problem for the focussing phase, i.e. we have models for both the flow of the fluid and the concentration of the particles. We end up here with the formulation of the focussing phase as an optimal control problem. In this work we will make use of the *discretize-then-optimize* approach, see [14]. In Chapter 5 we will discuss a spatial discretization for the optimal control problem and we will give proof of the existence of an optimal solution afterwards. We also present the optimality system which allows us to solve the optimization problem numerically.

5 Discretization in space

In this chapter, we will consider the spatial discretization of the optimal control problem (4.14)–(4.19). At first, we will review the finite element method. Then we will discretize the Stokes–Brinkman equation, the advection-diffusion equation and also the objective functional. Afterwards the optimality system for the discrete optimal control problem will be derived.

5.1 The finite element method

In this section, we review the finite element method. For simplicity, we restrict ourselves to problems on a two-dimensional domain $\Omega \subset \mathbb{R}^2$. The finite element method makes use of the variational formulation of a partial differential equation.

Let H be a Hilbert space and let H^* be its dual space. Let furthermore $a : H \times H \to \mathbb{R}$ denote a bilinear form which is supposed to be continuous and elliptic. Then we consider the following problem: For $f \in H^*$, find $u \in H$ such that

$$a(u, v) = f(v) \tag{5.1}$$

holds true for all $v \in H$.

Remark 5.1. A problem like (5.1) arises for example if the weak formulation of the Poisson equation [27] is derived. We use this weak formulation (5.1) to keep things simple as we want to focus on the basic finite element technique.

In general, H is an infinite dimensional Hilbert space. This makes it very difficult to solve the equation (5.1) for a given right-hand side $f \in H^*$. The basic idea of the finite element method is to use a spatial discretization of Ω and to solve (5.1) just in a finite dimensional number of points in Ω. Thus, for a finite dimensional function space $V_h \subset H$ we consider the problem: For $f \in H^*$, find $u_h \in V_h$ such that

$$a(u_h, v_h) = f(v_h) \tag{5.2}$$

holds true for all $v_h \in V_h$. Our goal is to find v_h such that v_h is a good approximation to the original solution u. Let $\mathcal{T}_h(\Omega)$ be a simplicial triangulation of Ω. Figure 5.1 shows one example of such a triangulation using triangles. The paramter h in $\mathcal{T}_h(\Omega)$ measures the fineness of the triangulation.

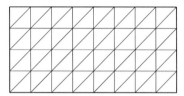

Figure 5.1: A simplicial triangulation of the domain.

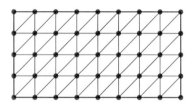

Figure 5.2: The degrees of freedom, here the vertices of the triangulation.

A finite element (K, P, Σ) consists of a simplex $K \subset \Omega$, a finite dimensional function space P and degrees of freedom Σ. For the number n of degrees of freedom, it is important that $\dim(P) = n$ holds true. Otherwise, the finite element is not well-defined. In our case, the simplex K is always a triangle. The function space P is the space of polynomials with degree 1 or 2, namely $\mathbb{P}_1[x]$ or $\mathbb{P}_2[x]$, and the degrees of freedom are the vertices of the triangle or the vertices and midpoints of the edges, respectively. Figure 5.2 shows an example for a choice of degrees of freedom. Note that P is chosen as the finite dimensional space V_h.

Now we define on each triangle n ansatz functions depending on the degrees of freedom. Then we approximate the solution u by the following ansatz

$$u(x) \approx u_h(x) := \sum_{i=1}^{n} \alpha_i \varphi_i(x) \tag{5.3}$$

where $\{\varphi_i\}_{i=1}^{n} \subset P$ denote the basis functions corresponding to the degrees of freedom. The coefficients $\alpha_i, 1 \leq i \leq n$, are unknown.

If we put (5.3) in (5.1) we arrive at the following problem: Find $\alpha_i, 1 \leq i \leq n$, such that

$$a(\sum_{i=1}^{n} \alpha_i \varphi_i, \varphi) = f(\varphi) \tag{5.4}$$

holds true for all $\varphi \in V_h$. Since V_h is finite dimensional, it is sufficient to fulfill the equations

$$a(\sum_{i=1}^{n} \alpha_i \varphi_i, \varphi_j) = f(\varphi_j), \quad j = 1, \ldots, n,$$

because $\{\varphi_i\}_{i=1}^n$ are the basis vectors of V_h. Due to the bilinearity of a we have

$$\sum_{i=1}^n \alpha_i a(\varphi_i, \varphi_j) = f(\varphi_j), \quad j = 1, \ldots, n,$$

and, therefore, we can rewrite the problem in the matrix form

$$A\boldsymbol{\alpha} = \mathbf{f},$$

where $A = [A_{ij}] \in \mathbb{R}^{n \times n}$ with $A_{ij} = a(\varphi_j, \varphi_i), \mathbf{f} = [f(\varphi_1), \ldots, f(\varphi_n)]^T \in \mathbb{R}^n$ and $\boldsymbol{\alpha} = [\alpha_1, \ldots, \alpha_n]^T \in \mathbb{R}^n$ is unknown. This problem can be solved easily now.

Implementation of the finite element method

We will give a brief review how this method is implemented. Therefore, we consider the special case that the bilinear form $a(\cdot, \cdot)$ is defined as

$$a(\varphi_i, \varphi_j) = \int_\Omega \nabla \varphi_i(x) \cdot \nabla \varphi_j(x) \, \mathrm{d}x. \qquad (5.5)$$

This expression suggests that we need to know the finite element basis functions $\{\varphi_i\}_{i=1}^n$. But there is an easier way to compute (5.5) without computing all the basis functions explicitly.

At first, we consider the reference triangle K_{ref} which is the left triangle shown in Figure 5.3. In this small example, we choose $P = \mathbb{P}_2[x]$ and the degrees of freedom are the vertices and midpoints of the edges of the reference triangle, $p_1 = (0,0), p_2 = (1,0),$ $p_3 = (0,1), p_4 = (\frac{1}{2}, 0), p_5 = (0, \frac{1}{2}), p_6 = (\frac{1}{2}, \frac{1}{2})$.

Thus, the six quadratic basis functions are

$$\begin{aligned}
\hat{\varphi}_1(x, y) &= (1 - x - y)(1 - 2x - 2y), \\
\hat{\varphi}_2(x, y) &= x(2x - 1), \\
\hat{\varphi}_3(x, y) &= y(2y - 1), \\
\hat{\varphi}_4(x, y) &= 4x(1 - x - y), \\
\hat{\varphi}_5(x, y) &= 4y(1 - x - y), \\
\hat{\varphi}_6(x, y) &= 4xy.
\end{aligned}$$

Note that $\hat{\varphi}_i(p_j) = \delta_{ij}$.

Figure 5.3 suggests that there exists an isomorphism between the reference triangle and an arbitrary triangle. We can compute this isomorphism $F : K_{\mathrm{ref}} \to K$ explicitly as

$$\begin{pmatrix} x \\ y \end{pmatrix} = F(\hat{x}, \hat{y}) = A \begin{pmatrix} \hat{x} \\ \hat{y} \end{pmatrix} + \mathbf{b},$$

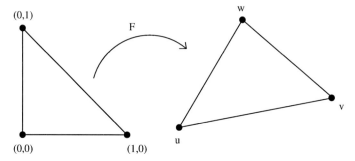

Figure 5.3: The isomoprhism between the reference triangle K_{ref} and an arbitrary triangle K.

where $A = [\mathbf{w} - \mathbf{u}, \mathbf{v} - \mathbf{u}] \in \mathbb{R}^{2 \times 2}$ and $\mathbf{b} = \mathbf{u} \in \mathbb{R}^2$. Here, x, y denote the coordinates of the arbitrary triangle K, $\hat{x}, \hat{y}$ denote the coordinates in the reference triangle K_{ref}, and $\mathbf{w}, \mathbf{v}, \mathbf{u}$ are the vertices of K.

The test function $\hat{\varphi}_i$ on the reference triangle corresponds to the degree of freedom p_i. Let φ_i be a test function on the arbitrary triangle, which corresponds to the point $F(p_i)$. Then we have the following relationship between $\hat{\varphi}_i, \varphi_i$,

$$\varphi_i \circ F = \hat{\varphi}_i,$$

which means we can compute φ_i as

$$\varphi_i = \hat{\varphi}_i \circ F^{-1},$$

where $F^{-1}(x, y) = A^{-1} \left(\begin{pmatrix} x \\ y \end{pmatrix} - \mathbf{b} \right)$.

For the derivative we use the chain rule

$$D\varphi_i = (D\hat{\varphi}_i \circ F^{-1})DF^{-1},$$

where $DF^{-1} = A^{-1}$. Thus, it holds

$$\nabla\varphi_i = A^{-T}(\nabla\hat{\varphi}_i \circ F^{-1}).$$

We can use the substitution rule $\int_{\phi(U)} f(v)\,\mathrm{d}v = \int_U f(\phi(u))|\det(D\phi)(u)|\,\mathrm{d}u$ for the integration and have

$$\int_K \nabla\varphi_i(x) \cdot \nabla\varphi_j(x)\,\mathrm{d}x = \int_{K_{\text{ref}}} (A^{-T}\nabla\hat{\varphi}_i) \cdot (A^{-T}\nabla\hat{\varphi}_j)|\det A|\,\mathrm{d}\hat{x},$$

where K is the arbitrary triangle, K_{ref} is the reference triangle and $F(K_{\text{ref}}) = K$.

5.2 Finite element discretization of the Stokes–Brinkman equation

In this section we will describe the spatial discretization of the Stokes–Brinkman equations (4.1)–(4.6) using the finite element method.

Therefore, we recall first the weak formulation from Section 4.1 for the Stokes–Brinkman equation: Find $(\mathbf{v}, p) \in V(0, T) \times P(0, T)$ such that

$$\frac{\mathrm{d}}{\mathrm{d}t} m_{sb}(\mathbf{v}, \boldsymbol{\varphi}) + a_{sb}(\mathbf{v}, \boldsymbol{\varphi}) - b_{sb}(\boldsymbol{\varphi}, p) = 0, \quad t \in (0, T], \tag{5.6}$$

$$b_{sb}(\mathbf{v}, \psi) = 0, \quad t \in (0, T], \tag{5.7}$$

$$(\mathbf{v}(\cdot, 0), \boldsymbol{\varphi})_{L^2(\Omega)} = (\mathbf{v}_0, \boldsymbol{\varphi})_{L^2(\Omega)} \tag{5.8}$$

holds true for all $\boldsymbol{\varphi} \in V$ and $\psi \in L^2(\Omega)$. The definition of the function spaces was given in Section 4.1.

Let $\mathcal{T}_h(\Omega)$ denote a simplicial triangulation of the domain Ω as shown in Figure 4.1. And let $\{p_i\}_{i=1}^{N_v}$ be the set of grid points if the triangulation. This set includes the vertices of all triangles of $\mathcal{T}_h(\Omega)$ as well as the midpoints of the edges.

For the spatial discretization, the finite dimensional subspaces $V_h \subset V(0, T)$ and $P_h \subset P(0, T)$ are required. Therefore, we use P2/P1-Taylor Hood finite elements for the discretization of the Stokes–Brinkman equation (5.6)–(5.8), i.e. quadratic elements for the discretization of the velocity and linear elements for the pressure. The finite dimensional ansatz spaces are given by

$$V_h := \{\mathbf{v}_h \in C(\bar{\Omega}) \,|\, \mathbf{v}_h\,|_K \in P_2(K)^2 \text{ for all } K \in \mathcal{T}_h(\Omega), \mathbf{v}_h\,|_{\Gamma_{\text{lat}} \cup \Gamma_{\text{in}}} = 0\},$$

$$P_h := \{p_h \in C(\bar{\Omega}) \,|\, p_h\,|_K \in P_1(K) \text{ for all } K \in \mathcal{T}_h(\Omega)\}.$$

Furthermore, let $\mathbf{v}_{h,\text{in}}^{(i)}, 1 \leq i \leq 2$, denote the L^2-projection of the given inflow functions $\mathbf{v}_{\text{in}}^{(i)}$.

Remark 5.2. Note that, the finite element ansatz spaces for Stokes-type systems have to fullfill the Ladyschenskaja-Babuska-Brezzi condition [40]. This condition is necessary and sufficient for the well posedness of the problem. For more detail, we refer to [40]. This claim is fulfilled by P2/P1-Taylor-Hood finite elements, see [31].

Using this framework, the problem reads:

Compute $(\mathbf{v}_h, p_h) \in H^1(0, T; V_h) \times L^2(0, T; P_h)$ such that

$$\frac{\mathrm{d}}{\mathrm{d}t} m_{sb}(\mathbf{v}_h, \boldsymbol{\varphi}_h) + a_{sb}(\mathbf{v}_h, \boldsymbol{\varphi}_h) - b_{sb}(\boldsymbol{\varphi}_h, p_h) = 0, \quad t \in (0, T], \tag{5.9}$$

$$b_{sb}(\mathbf{v}_h, \psi_h) = 0, \quad t \in (0, T], \tag{5.10}$$

$$(\mathbf{v}_h(\cdot, 0), \boldsymbol{\varphi}_h)_{L^2(\Omega)} = (\mathbf{v}_0, \boldsymbol{\varphi}_h)_{L^2(\Omega)} \tag{5.11}$$

holds for all $\boldsymbol{\varphi}_h \in V_h, \psi_h \in P_h$.

Let $(\varphi_i)_{i=1}^{N_v}$ and $(\psi_j)_{j=1}^{N_p}$ denote the bases of V_h and P_h, respectively. Then we start with the following ansatz

$$\mathbf{v}(x,t) \approx \sum_{i=1}^{N_v} \alpha_i(t)\varphi_i(x) =: \mathbf{v}_h(x,t),$$

$$p(x,t) \approx \sum_{j=1}^{N_p} \beta_j(t)\psi_j(x) =: p_h(x,t),$$

that means the unknowns are the time-dependent coefficient vectors
$\boldsymbol{\alpha}(t) = (\alpha_1(t), \ldots, \alpha_{N_v}(t))^T \in \mathbb{R}^{N_v}$ and $\boldsymbol{\beta}(t) = (\beta_1(t), \ldots, \beta_{N_p}(t))^T \in \mathbb{R}^{N_p}$.

The inflow function $\mathbf{v}_{h,\text{in}}^{(i)}$ on the related inflow boundary is defined as

$$\mathbf{v}_{h,\text{in}}^{(i)}(x,t) = \begin{pmatrix} 0 \\ -u_i(t)(x_1 - b_1)(b_2 - x_1) \end{pmatrix} =: u_i(t)\mathbf{w}^{(i)}(x), \quad i = 1,2,$$

with $\mathbf{w}^{(i)}(x) = \begin{pmatrix} 0 \\ -(x_1 - b_1)(b_2 - x_1) \end{pmatrix}$ for $x = [x_1, x_2]^T \in \Gamma_{\text{in}}^{(i)}$ and $t \in [0,T]$ and $\mathbf{v}_{h,\text{in}}^{(i)}(x,t) = 0$ for $x \in \Omega \setminus \Gamma_{\text{in}}^{(i)}$, $i = 1,2$ and $t \in [0,T]$.

Let now $\{\varphi_i\}_{i=1}^{n_v}$ denote the set of basis functions corresponding to all grid points of the mesh where no Dirichlet boundary condition is posed. Then, using the remaining basis functions $\{\varphi_i\}_{i=n_v+1}^{N_v}$, the semi-discretized solution $\mathbf{v}_h$ of the Stokes–Brinkman equation (5.9)–(5.11) can be expressed as

$$\mathbf{v}_h(x,t) = \sum_{i=1}^{n_v} \alpha_i(t)\varphi_i(x) + \sum_{i=n_v+1}^{N_v} \alpha_i(t)\varphi_i(x) =: \mathbf{v}_1(x,t) + \mathbf{v}_2(x,t).$$

Due to the Dirichlet conditions, namely the no-slip condition (4.4) and the inflow conditions (4.3), the coefficients $\alpha_i(t)$ for $i = n_v + 1, \ldots, N_v$ are known. The coefficients corresponding to the no-slip boundary are equal to zero, whereas the coefficients corresponding to the inflow boundary can be computed explicitly using $\mathbf{v}_{h,\text{in}}^{(i)}$. This means that $\mathbf{v}_2(x,t) = u_1(t)\sum_r \mathbf{w}^{(1)}(p_r)\varphi_r(x) + u_2(t)\sum_k \mathbf{w}^{(2)}(p_k)\varphi_k(x)$. Putting this ansatz in (5.9)–(5.11), the equations

$$\sum_{i=1}^{n_v} \frac{\mathrm{d}}{\mathrm{d}t}\alpha_i(t)m_{sb}(\varphi_i, \varphi_j) + \sum_{i=1}^{n_v} \alpha_i(t)a_{sb}(\varphi_i, \varphi_j) - \sum_{i=1}^{N_p} \beta_i b_{sb}(\varphi_j, \psi_i) = -\frac{\mathrm{d}}{\mathrm{d}t}m_{sb}(\mathbf{v}_2, \varphi_j) - a_{sb}(\mathbf{v}_2, \varphi_j),$$

$$\sum_{i=1}^{n_v} \alpha_i(t)b_{sb}(\varphi_i, \psi_j) = -b_{sb}(\mathbf{v}_2, \psi_j),$$

$$\sum_{i=1}^{n_v} \alpha_i(0)(\varphi_i, \varphi_j)_{L^2(\Omega)} = (\mathbf{v}_0, \varphi_j)_{L^2(\Omega)} - m_{sb}(\mathbf{v}_2, \varphi_j),$$

have to be fulfilled for all $j = 1, \ldots, n_v$.

Remark 5.3. We test only with the basis functions $\{\boldsymbol{\varphi}_i\}_{i=1}^{n_v}$, because the set of functions $\{\boldsymbol{\varphi}_i\}_{i=n_v+1}^{N_v}$ are not elements of V_h as they are not zero on $\Gamma_{\text{lat}} \cup \Gamma_{\text{in}}$.

These equations lead to a system of differential-algebraic equations (DAEs)

$$M\frac{\mathrm{d}}{\mathrm{d}t}\mathbf{v}(t) = A\mathbf{v}(t) + B^T\mathbf{p}(t) + F_0\mathbf{u}(t) + F_1\frac{\mathrm{d}}{\mathrm{d}t}\mathbf{u}(t), \tag{5.12}$$

$$0 = B\mathbf{v}(t) + L\mathbf{u}(t), \tag{5.13}$$

$$M\mathbf{v}(0) = \mathbf{v}_{0,h}, \tag{5.14}$$

where $\mathbf{v}(t) = [\alpha_1(t), \ldots, \alpha_{n_v}(t)]^T \in \mathbb{R}^{n_v}$, $\mathbf{p}(t) = [\beta_1(t), \ldots, \beta_{N_p}(t)]^T \in \mathbb{R}^{N_p}$, $\mathbf{u}(t) = [u_1(t), u_2(t)]^T$ and the entries of the coefficient matrices are given by

$$M_{i,j} = \rho \int_\Omega \boldsymbol{\varphi}_i \cdot \boldsymbol{\varphi}_j \, \mathrm{d}x, \qquad\qquad i,j = 1,\ldots,n_v,$$

$$A_{i,j} = -\nu \int_\Omega \nabla\boldsymbol{\varphi}_i : \nabla\boldsymbol{\varphi}_j \, \mathrm{d}x - \nu K^{-1} \int_\Omega \chi_{\Omega_2} \boldsymbol{\varphi}_i \cdot \boldsymbol{\varphi}_j \, \mathrm{d}x, \qquad\qquad i,j = 1,\ldots,n_v,$$

$$B_{i,j} = \int_\Omega (\nabla \cdot \boldsymbol{\varphi}_j)\psi_i \, \mathrm{d}x, \qquad\qquad i = 1,\ldots,N_p,\, j = 1,\ldots,n_v,$$

$$(F_0)_{i,j} = -\nu \int_\Omega \nabla(\sum_r \mathbf{w}^{(j)}(p_r) \cdot \boldsymbol{\varphi}_r) \cdot \nabla\boldsymbol{\varphi}_i \, \mathrm{d}s, \qquad\qquad i = 1,\ldots,n_v,\, j = 1,2,$$

$$(F_1)_{i,j} = -\rho \int_\Omega (\sum_r \mathbf{w}^{(j)}(p_r) \cdot \boldsymbol{\varphi}_r) \cdot \boldsymbol{\varphi}_i \, \mathrm{d}s, \qquad\qquad i = 1,\ldots,n_v,\, j = 1,2,$$

$$L_{i,j} = \int_\Omega \psi_i \nabla \cdot (\sum_r \mathbf{w}^{(j)}(p_r)\boldsymbol{\varphi}_r) \, \mathrm{d}s, \qquad\qquad i = 1,\ldots,n_v,\, j = 1,2.$$

The vector $\mathbf{v}_{0,h}$ gathers the contribution of $\mathbf{v}_0$ and of $\mathbf{v}_2$. Taking into account that the velocity vector corresponding to the subdomain Ω_1 is required in the advection-diffusion equation (4.11)–(4.13), we add the output equation $\mathbf{z}(t) = C_1\mathbf{v}(t) \in \mathbb{R}^{n_{v,1}}$ with $C_1 \in \mathbb{R}^{n_{v,1}\times n_v}$ to system (5.12)–(5.14). As the result we obtain the DAE control system

$$E\frac{\mathrm{d}}{\mathrm{d}t}\mathbf{x}(t) = S\mathbf{x}(t) + F\hat{\mathbf{u}}(t), \quad E\mathbf{x}(0) = \mathbf{x}_0, \tag{5.15}$$

$$\mathbf{z}(t) = C_{\text{out}}\mathbf{x}(t), \tag{5.16}$$

where $E = \begin{bmatrix} M & 0 \\ 0 & 0 \end{bmatrix}, S = \begin{bmatrix} A & B^T \\ B & 0 \end{bmatrix}, F = \begin{bmatrix} F_0 & F_1 \\ L & 0 \end{bmatrix}, C_{\text{out}} = \begin{bmatrix} C_1 & 0 \end{bmatrix}$. Furthermore, $x(t) = [\mathbf{v}^T(t), \mathbf{p}^T(t)]^T \in \mathbb{R}^{n_v+N_p}$ is the state vector, $\mathbf{x}_0 = [\mathbf{v}_{0,h}^T, \mathbf{0}]$ is the initial vector, $\hat{\mathbf{u}}(t) = [\mathbf{u}^T(t), \frac{\mathrm{d}}{\mathrm{d}t}\mathbf{u}^T(t)]^T \in \mathbb{R}^4$ is the input and $\mathbf{z}(t) \in \mathbb{R}^{n_{v,1}}$ is the output. To keep notation clean, we define $n_p := N_p$ for the dimension of the pressure.

5.3 Finite element discretization of the advection-diffusion equation

In this section, we discuss the spatial discretization of the advection-diffusion equation (4.11)–(4.13) using the stabilized Petrov-Galerkin (SUPG) method [16] based on P2 finite elements.

We consider the weak formulation for the advection-diffusion equation (4.11)–(4.13) given in Section 4.2. Let $\mathcal{T}_h(\Omega_1)$ be a simplicial triangulation of Ω_1. Denote by

$$V_{2,h} := \{c_h \in C(\bar{\Omega}) \mid c_h \mid_K \in P_2(K)^2 \text{ for all } K \in \mathcal{T}_h(\Omega_1)\},$$

a finite dimensional subspace of V_2 with the basis $\{\varphi_i\}_{i=1}^{n_c}$. We use the ansatz

$$c(x,t) \approx \sum_{i=1}^{n_c} \gamma_i(t)\varphi_i(x) =: c_h(x,t), \qquad (5.17)$$

where the coefficient vector $\mathbf{c}(t) = (\gamma_1(t), \ldots, \gamma_{n_c}(t))^T \in \mathbb{R}^{n_c}$ has to be determined. For an increase of stability, we use the SUPG method which was originally introduced in [16]. This method was already applied to optimal control problems, for example, in [19]. Briefly speaking, the idea of this method is to consider a perturbation of the original test functions $\varphi_j + \zeta_j$ and use them as the test functions in the weak formulation instead. This ansatz leads to

$$\frac{\mathrm{d}}{\mathrm{d}t} m_{ad}(c_h, \varphi_j + \zeta_j) + a_{ad}(c_h, \varphi_j + \zeta_j) = 0, \quad t \in (0,T],$$

$$(c_h(\cdot, 0), \varphi_j)_{L^2(\Omega_1)} = (c_0, \varphi_j)_{L^2(\Omega_1)}.$$

In [16], it was suggested to choose

$$\zeta_j|_K = \tau_K h_K \mathbf{v}_h \cdot \nabla\varphi_j, \quad K \in \mathcal{T}_h(\Omega_1),$$

where $\tau_K > 0$ denotes a stabilization parameter, and h_K the diameter of the triangle K. Using the perturbed test functions, we add an additional element-wise diffusion for more stability. For more details, see [16].

This leads to the final formulation: find $c_h \in C^1(0, T; V_h^2)$ such that

$$\sum_{K \in \mathcal{T}_h(\Omega_1)} \frac{\mathrm{d}}{\mathrm{d}t}(c_h, \varphi + \tau_K h_K \mathbf{v}_h \cdot \nabla\varphi)_{L^2(K)} + a_{ad}(c_h, \varphi)$$

$$+ \sum_{K \in \mathcal{T}_h(\Omega_1)} \tau_K h_K (-D\Delta c_h + \mathbf{v}_h \cdot \nabla c_h, \mathbf{v}_h \cdot \nabla\varphi)_{L^2(K)} = 0, \qquad (5.18)$$

$$(c_h(\cdot, 0), \varphi)_{L^2(\Omega_1)} = (c_0, \varphi)_{L^2(\Omega_1)},$$

holds true for all $\varphi \in V_{2,h}$.

Taking $\varphi = \varphi_i, i = 1, \ldots, n_c$, we end up with the linear time-varying system of ordinary differential equations

$$(M_1 + M_2(\mathbf{v}(t)))\frac{\mathrm{d}}{\mathrm{d}t}\mathbf{c}(t) + A_1(\mathbf{v}(t))\mathbf{c}(t) = 0, \quad t \in (0, T),$$

$$M_1\mathbf{c}(0) = \mathbf{c}_{0,h},$$

where the matrices are given by

$$(A_1(\mathbf{v}(t)))_{i,j} = \int_{\Omega_1} D\nabla\varphi_j \cdot \nabla\varphi_i \, \mathrm{d}x$$

$$- \sum_{k=1}^{n_v} \alpha_k(t) \int_{\Omega_1} \boldsymbol{\varphi}_k \cdot \nabla\varphi_i \varphi_j \, \mathrm{d}x$$

$$- \sum_{k=1}^{n_v} \alpha_k(t) \sum_{K \in \mathcal{T}_h(\Omega_1)} \tau_K h_K \int_K \Delta\varphi_j \nabla\varphi_i \cdot \boldsymbol{\varphi}_k \, \mathrm{d}x$$

$$+ \sum_{k_1,k_2=1}^{n_v} \alpha_{k_1}(t)\alpha_{k_2}(t) \sum_{K \in \mathcal{T}_h(\Omega_1)} \tau_K h_K \int_K \nabla\varphi_j \cdot \boldsymbol{\varphi}_{k_1} \nabla\varphi_i \cdot \boldsymbol{\varphi}_{k_2} \, \mathrm{d}x,$$

and furthermore

$$(M_1)_{i,j} = \int_{\Omega_1} \varphi_i\varphi_j \, \mathrm{d}x,$$

$$(M_2(\mathbf{v}(t)))_{i,j} = \sum_{k=1}^{n_v} \alpha_k(t) \sum_{K \in \mathcal{T}_h(\Omega_1)} \tau_K h_K \int_K \varphi_j\boldsymbol{\varphi}_k \cdot \nabla\varphi_i \, \mathrm{d}x,$$

where $\mathbf{v}(t) = [\alpha_1(t), \ldots, \alpha_{n_v}(t)]^T$ is the semi-discretized velocity vector satisfying the Stokes–Brinkman equation (5.12)–(5.14).

5.4 Spatial discretization of the optimal control problem

In this section, finally the functional (4.14) is discretized using finite elements.

Recall the functional (4.14) and consider the first summand $\frac{1}{2}\|c(T) - c^{\mathrm{foc}}\|_{L^2(\Omega_1)}^2$. We use the approximation $c_h(x, t)$ from (5.17) for $c(x, t)$ and the ansatz

$$c^{\mathrm{foc}}(x) \approx \sum_{i=1}^{n_c} \eta_i\varphi_i(x) =: c_h^{\mathrm{foc}}(x),$$

where the coefficients $\eta_i, i = 1, \ldots, n_c$, are known, because c^{foc} is given. Thus, we have

$$\int_{\Omega_1} (c_h(x,T) - c_h^{\text{foc}}(x))^2 \, \mathrm{d}x = \int_{\Omega_1} \left(\sum_{i=1}^{n_c} (\gamma_i(T) - \eta_i) \varphi_i \right) \cdot \left(\sum_{j=1}^{n_c} (\gamma_j(T) - \eta_j) \varphi_j \right) \, \mathrm{d}x$$

$$= \int_{\Omega_1} \sum_{i=1}^{n_c} \left((\gamma_i(T) - \eta_i) \sum_{j=1}^{n_c} (\gamma_j(T) - \eta_j) \varphi_j \varphi_i \right) \, \mathrm{d}x$$

$$= \sum_{i=1}^{n_c} \left((\gamma_i(T) - \eta_i) \sum_{j=1}^{n_c} \left(\int_{\Omega_1} \varphi_j \varphi_i \, \mathrm{d}x \right) (\gamma_j(T) - \eta_j) \right)$$

$$= (\mathbf{c}(T) - \mathbf{c}^{\text{foc}}, M_1(\mathbf{c}(T) - \mathbf{c}^{\text{foc}}))_2,$$

$$= \|\mathbf{c}(T) - \mathbf{c}^{\text{foc}}\|_{M_1}^2,$$

where $\mathbf{c}(T) = [\gamma_1(T), \ldots, \gamma_{n_c}(T)]^T$, $\mathbf{c}^{\text{foc}} = [\eta_1, \ldots, \eta_{n_c}]^T$.

The semi-discretized optimization problem reads then

$$\text{Minimize } J(\mathbf{u}) := \frac{1}{2} \|\mathbf{c}(T) - \mathbf{c}^{\text{foc}}\|_{M_1}^2 + \frac{\sigma}{2} \|\mathbf{u}\|_{H^3(0;T,\mathbb{R}^2)}^2 \tag{5.19}$$

such that the equations

$$M \frac{\mathrm{d}}{\mathrm{d}t} \mathbf{v}(t) = A\mathbf{v}(t) + B^T \mathbf{p}(t) + F_0 \mathbf{u}(t) + F_1 \frac{\mathrm{d}}{\mathrm{d}t} \mathbf{u}(t), \tag{5.20}$$

$$0 = B\mathbf{v}(t) + L\mathbf{u}(t), \tag{5.21}$$

$$M\mathbf{v}(0) = \mathbf{v}_{0,h} \tag{5.22}$$

and

$$(M_1 + M_2(\mathbf{v}(t))) \frac{\mathrm{d}}{\mathrm{d}t} \mathbf{c}(t) + A_1(\mathbf{v}(t))\mathbf{c}(t) = 0 \tag{5.23}$$

$$M_1 \mathbf{c}(0) = \mathbf{c}_{0,h}, \tag{5.24}$$

hold true.

6 Existence of a minimizer

In this chapter, we recall the previous optimization problem and show the existence of a minimizer. Therefore, we investigate the solvability of the semi-discretized Stokes–Brinkman and advection-diffusion equations.

The optimization problem reads

$$\text{Minimize } J(\mathbf{u}) = \frac{1}{2}\|\mathbf{c}(T) - \mathbf{c}^{\text{foc}}\|_{M_1}^2 + \frac{\sigma}{2}\|\mathbf{u}\|_{H^3}^2, \tag{6.1}$$

where

$$M\frac{\mathrm{d}}{\mathrm{d}t}\mathbf{v}(t) = A\mathbf{v}(t) + B^T\mathbf{p}(t) + F_0\mathbf{u}(t) + F_1\frac{\mathrm{d}}{\mathrm{d}t}\mathbf{u}(t), \tag{6.2}$$

$$\mathbf{0} = B\mathbf{v}(t) + L\mathbf{u}(t), \tag{6.3}$$

$$M\mathbf{v}(0) = \mathbf{v}_0, \tag{6.4}$$

and

$$(M_1 + M_2(\mathbf{v}(t)))\frac{\mathrm{d}}{\mathrm{d}t}\mathbf{c}(t) + A_1(\mathbf{v}(t))\mathbf{c}(t) = \mathbf{0}, \tag{6.5}$$

$$M_1\mathbf{c}(0) = \mathbf{c}_0. \tag{6.6}$$

Remark 6.1. The matrices $(M_1 + M_2(\mathbf{v}(t)))$ and M_1 are in our numerical experiments always nonsingular. Thus, for a simplicity of notation, we consider the equation

$$\frac{\mathrm{d}}{\mathrm{d}t}\mathbf{y}(t) = A(\mathbf{v}(t))\mathbf{y}(t),$$

$$\mathbf{y}(0) = \mathbf{y}_0,$$

instead of (6.5)–(6.6). Note that (6.5)–(6.6) can easily be transformed into this form multiplying with $(M_1 + M_2(\mathbf{v}))^{-1}$ and M_1^{-1}, respectively.

We introduce now the solution operators $\mathcal{S}_1 : C^1(0, T; \mathbb{R}^4) \to H^1(0, T; \mathbb{R}^{n_v})$ and $\mathcal{S}_2 : C^0(0, T; \mathbb{R}^{n_v}) \to C^1(0, T; \mathbb{R}^n)$. Here, $\mathbf{v} = \mathcal{S}_1(\mathbf{w})$ for $\mathbf{w} \in C^1(0, T; \mathbb{R}^4)$ with

$$E\frac{\mathrm{d}}{\mathrm{d}t}\mathbf{x}(t) = S\mathbf{x}(t) + F\mathbf{w}(t), \tag{6.7}$$

$$\mathbf{v}(0) = -P_\mathbf{v}(S^{-1}F\mathbf{w}(0)), \tag{6.8}$$

where $E = \begin{bmatrix} M & 0 \\ 0 & 0 \end{bmatrix}$, $S = \begin{bmatrix} A & B^T \\ B & 0 \end{bmatrix}$, $F = [\bar{F}_1^T, \bar{F}_2^T]^T$, $\bar{F}_1 = [F_0 \ F_1]$, $\bar{F}_2 = [L \ 0]$,

$\mathbf{x} = [\mathbf{v}^T, \mathbf{p}^T]^T$ and $P_{\mathbf{v}} = [I, 0]$ denotes the selection matrix for the first component with $P_{\mathbf{v}}\mathbf{x} = \mathbf{v}$. For $\mathcal{S}_2(\mathbf{v}) = \mathbf{y}$ it holds

$$\frac{\mathrm{d}}{\mathrm{d}t}\mathbf{y}(t) = A(\mathbf{v}(t))\mathbf{y}(t), \tag{6.9}$$

$$\mathbf{y}(0) = \mathbf{y}_0. \tag{6.10}$$

Remark 6.2. Condition (6.8) implies that the stationary solution is used as initial condition. We will explain later the reason why this makes a good choice.

Furthermore, let $\mathcal{I}\colon H^2(0,T;\mathbb{R}^4) \to C^1(0,T;\mathbb{R}^4)$, $\mathcal{L}\colon H^1(0,T;\mathbb{R}^{n_v}) \to C^0(0,T;\mathbb{R}^{n_v})$ be the compact embedding operators and let $\mathcal{J} : H^3(0,T;\mathbb{R}^2) \to H^2(0,T;\mathbb{R}^4)$ be defined as

$$\mathcal{J}(\mathbf{u}) = \begin{pmatrix} \mathbf{u} \\ \frac{\mathrm{d}}{\mathrm{d}t}\mathbf{u} \end{pmatrix}.$$

Then the cost functional (6.1) can be written as

$$J(\mathbf{u}) = \frac{1}{2}\|\mathcal{S}_2(\mathcal{L}(\mathcal{S}_1(\mathcal{I}(\mathcal{J}(\mathbf{u})))))(T) - \mathbf{c}^{\text{foc}}\|_{M_1}^2 + \frac{\sigma}{2}\|\mathbf{u}\|_{H^3}^2. \tag{6.11}$$

The aim of the optimization problem is to minimize the distance between the concentration of the particles at the final time T and a desired concentration profile. Note that, we have to add a regularization term $\frac{\sigma}{2}\|\mathbf{u}\|_{H^3}^2$ in order to show existence of a minimizer and to establish error estimates.

The next lemma provides the properties of the mapping $\mathcal{J}$.

Lemma 6.3. *The mapping $\mathcal{J}$ is linear and continuous.*

Proof. The linearity of $\mathcal{J}$ is obvious. The continuity of $\mathcal{J}$ immediately follows from the following estimate:

$$\|\mathcal{J}(\mathbf{u})\|_{H^2}^2 = \left\|\begin{pmatrix} \mathbf{u} \\ \frac{\mathrm{d}}{\mathrm{d}t}\mathbf{u} \end{pmatrix}\right\|_{H^2}^2 \leq \|\mathbf{u}\|_{H^2}^2 + \|\frac{\mathrm{d}}{\mathrm{d}t}\mathbf{u}\|_{H^2}^2 \leq 2\|\mathbf{u}\|_{H^3}^2. \qquad \square$$

6.1 Solvability of the semi-discretized Stokes–Brinkman equation

Next, we study the well-posedness of the Stokes–Brinkman equation (6.7)–(6.8).

Existence of solutions of Stokes-type equations is already well-studied in the literature [1, 43]. For our purpose, we will use [1] and especially the error estimates which were established there.

Theorem 6.4. *Let* $A, M \in \mathbb{R}^{n \times n}$ *and* $B \in \mathbb{R}^{m \times n}$ *with* $m \leq n$, *such that:*

- M *is symmetric positive definite;*
- $-A$ *is positive definite (not necessarily symmetric) on* $ker(B)$, *i. e. there exists a constant* $\alpha > 0$ *such that*

$$-\mathbf{v}^T A \mathbf{v} \geq \alpha \|\mathbf{v}\|^2 \text{ for all } \mathbf{v} \in ker(B);$$

- B *has full row rank* m.

Consider the initial value problem

$$E \frac{d}{dt} \begin{pmatrix} \mathbf{v} \\ \mathbf{p} \end{pmatrix} - S \begin{pmatrix} \mathbf{v} \\ \mathbf{p} \end{pmatrix} = \begin{pmatrix} \mathbf{g}_1 \\ \mathbf{g}_2 \end{pmatrix}, \tag{6.12}$$

$$M \mathbf{v}(0) = \mathbf{v}_0, \tag{6.13}$$

where $E = \begin{pmatrix} M & 0 \\ 0 & 0 \end{pmatrix}, S = \begin{pmatrix} A & B^T \\ B & 0 \end{pmatrix}, \mathbf{g}_1 \in C(0, T; \mathbb{R}^n), \mathbf{g}_2 \in C^1(0, T; \mathbb{R}^m)$ *and* $\mathbf{v}_0$ *satisfies*

$$BM^{-1}\mathbf{v}_0 + \mathbf{g}_2(0) = 0. \tag{6.14}$$

Then the initial value problem (6.12)–(6.13) *has a unique solution*

$$(\mathbf{v}, \mathbf{p}) \in C^1(0, T; \mathbb{R}^n) \times C^0(0, T; \mathbb{R}^m)$$

and there exist constants $c_1 \geq 0$ *and* $c_2 \geq 0$ *depending only on* $A, B,$ *and* M *such that*

$$\|\mathbf{v}\|_{L^2} \leq c_1 \|\mathbf{v}_0\| + c_2 (\|\mathbf{g}_1\|_{L^2} + \|\mathbf{g}_2\|_{L^2}), \tag{6.15}$$

$$\|\mathbf{p}\|_{L^2} \leq c_1 \|\mathbf{v}_0\| + c_2 (\|\mathbf{g}_1\|_{L^2} + \|\mathbf{g}_2\|_{L^2} + \|\frac{d}{dt}\mathbf{g}_2\|_{L^2}). \tag{6.16}$$

Proof. We begin with the existence and uniqueness of a solution. Therefore, we discuss a transformation of the Stokes–Brinkman equation from a DAE into an ordinary differential equation. The transformation of the Stokes–Brinkman equation will also become important in the use of the Lagrange framework and when we ask for reduced order models. Such a transformation was introduced in [43, 1].

We start with the original equations (6.12)–(6.13)

$$M \frac{d}{dt} \mathbf{v}(t) - A \mathbf{v}(t) - B^T \mathbf{p}(t) = \mathbf{g}_1(t) \tag{6.17}$$

$$-B \mathbf{v}(t) = \mathbf{g}_2(t), \tag{6.18}$$

$$M \mathbf{v}(0) = \mathbf{v}_0, \tag{6.19}$$

where $\mathbf{v}(t) \in \mathbb{R}^n, \mathbf{p}(t) \in \mathbb{R}^m$ are to be found and $\mathbf{g}_1 \in C^0(0, T; \mathbb{R}^n), \mathbf{g}_2 \in C^1(0, T; \mathbb{R}^m)$ are given.

We will decompose $\mathbf{v}$ in a clever way and consider a projection onto the kernel of BM^{-1}. Therefore, we define a projection matrix

$$\Pi := I - B^T(BM^{-1}B^T)^{-1}BM^{-1} \in \mathbb{R}^{n \times n}$$

onto $\ker(BM^{-1})$ along $\mathrm{im}(B^T)$. Note that the matrix $BM^{-1}B^T$ is invertible, since M is positive definite and B is of full row rank. It can be easily verified that $\Pi^2 = \Pi$ and $\Pi M = M\Pi^T$. Furthermore, the identities $\mathrm{im}(\Pi^T) = \ker(B)$ and $\ker(\Pi^T) = \mathrm{im}(M^{-1}B^T)$ can be shown.

Since Π is a projector, also Π^T is a projector. Then a potential solution $\mathbf{v}$ of the equations (6.12)–(6.13) can be additively decomposed as

$$\mathbf{v}(t) = \Pi^T\mathbf{v}(t) + (\mathbf{v}(t) - \Pi^T\mathbf{v}(t)) =: \mathbf{v}_1(t) + \mathbf{v}_2(t). \tag{6.20}$$

where

$$\mathbf{v}_1(t) = \Pi^T\mathbf{v}(t) = \Pi^T\mathbf{v}_1(t) \tag{6.21}$$

and

$$\begin{aligned}
\mathbf{v}_2(t) &= \mathbf{v}(t) - \Pi^T\mathbf{v}(t) = M^{-1}B^T(BM^{-1}B^T)^{-1}B\mathbf{v}(t) \\
&= -M^{-1}B^T(BM^{-1}B^T)^{-1}\mathbf{g}_2(t).
\end{aligned} \tag{6.22}$$

Putting the decomposition (6.20) into (6.17)–(6.19) we obtain

$$\begin{aligned}
M\frac{\mathrm{d}}{\mathrm{d}t}\mathbf{v}_1(t) - A\mathbf{v}_1(t) - B^T\mathbf{p}(t) &= \mathbf{g}_1(t) - AM^{-1}B^T(BM^{-1}B^T)^{-1}\mathbf{g}_2(t) \\
&\quad + B^T(BM^{-1}B^T)^{-1}\frac{\mathrm{d}}{\mathrm{d}t}\mathbf{g}_2(t)
\end{aligned} \tag{6.23}$$

$$B\mathbf{v}_1(t) = \mathbf{0}, \tag{6.24}$$

$$M\mathbf{v}_1(0) = \Pi\mathbf{v}_0. \tag{6.25}$$

Multiplying equation (6.23) from the left with $(BM^{-1}B^T)^{-1}BM^{-1}$ and using (6.24) we obtain the explicit expression for the pressure

$$\begin{aligned}
\mathbf{p}(t) &= -(BM^{-1}B^T)^{-1}BM^{-1}A\mathbf{v}_1(t) \\
&\quad - (BM^{-1}B^T)^{-1}BM^{-1}(\mathbf{g}_1(t) - AM^{-1}B^T(BM^{-1}B^T)^{-1}\mathbf{g}_2(t)) \\
&\quad - (BM^{-1}B^T)^{-1}\frac{\mathrm{d}}{\mathrm{d}t}\mathbf{g}_2(t).
\end{aligned} \tag{6.26}$$

Multiplying the equations (6.23) and (6.25) from the left with Π and using (6.21) and (6.26) we get the following DAE system

$$\Pi M\Pi^T\frac{\mathrm{d}}{\mathrm{d}t}\mathbf{v}_1(t) - \Pi A\Pi^T\mathbf{v}_1(t) = \Pi(\mathbf{g}_1(t) - AM^{-1}B^T(BM^{-1}B^T)^{-1}\mathbf{g}_2(t)), \tag{6.27}$$

$$\Pi^T\mathbf{v}_1(t) = \mathbf{v}_1(t), \tag{6.28}$$

$$\Pi M\Pi^T\mathbf{v}_1(0) = \Pi\mathbf{v}_0. \tag{6.29}$$

The system (6.27)–(6.29) is, together with equation (6.26), equivalent to the system (6.23)–(6.25) for $\mathbf{v}_1$. On the one hand a solution of (6.23)–(6.25) gives rise to a solution of (6.27)–(6.29) by setting $\mathbf{v}_1(t) = \Pi^T \mathbf{v}(t)$. On the other hand, any solution $\mathbf{v}_1$ of (6.27)–(6.29) is also a solution of (6.23)–(6.25) since the term $\Pi M \Pi^T \frac{d}{dt} \mathbf{v}_1(t)$ on the left side of the differential equation (7.15) can be written as $M \frac{d}{dt} \mathbf{v}_1(t)$ due to $\Pi M = M \Pi^T$ and $\Pi^T \mathbf{v}_1(t) = \mathbf{v}_1(t)$, and the matrix M is invertible. The initial condition for $\mathbf{v}$ is fulfilled since

$$
\begin{aligned}
M\mathbf{v}(0) &= M(\mathbf{v}_1(0) + \mathbf{v}_2(0)) = M(\Pi^T \Pi^T \mathbf{v}_1(0) + \mathbf{v}_2(0)) \\
&= \Pi M \Pi^T \mathbf{v}_1(0) + M\mathbf{v}_2(0) = \Pi \mathbf{v}_0 + M\mathbf{v}_2(0) \\
&= \mathbf{v}_0 - B^T(BM^{-1}B^T)^{-1}BM^{-1}\mathbf{v}_0 - B^T(BM^{-1}B^T)^{-1}\mathbf{g}_2(0) \\
&= \mathbf{v}_0 - B^T(BM^{-1}B^T)^{-1}(BM^{-1}\mathbf{v}_0 + \mathbf{g}_2(0)) \\
&= \mathbf{v}_0.
\end{aligned}
$$

The last step is eliminating the algebraic condition $\Pi^T \mathbf{v}_1(t) = \mathbf{v}_1(t)$. We can factor the projector matrix Π as

$$
\Pi = \Theta_l \Theta_r^T, \text{ where } \Theta_l, \Theta_r \in \mathbb{R}^{n \times (n-m)} \text{ and }
$$
$$
\Theta_l^T \Theta_r = I.
$$

This factorization is not unique; it can be obtained by considering representing matrices of a factorization of the linear map associated to Π as a surjection followed by an injection: $\Pi = \Theta_l \Theta_r^T$. In the equation $L_{\Theta_l} \circ (L_{\Theta_r^T} \circ L_{\Theta_l}) \circ L_{\Theta_r^T} = L_\Pi \circ L_\Pi = L_\Pi = L_{\Theta_l} \circ \mathrm{id} \circ L_{\Theta_r^T}$ we may cancel L_{Θ_l} (since it is injective) and $L_{\Theta_r^T}$ (since it is surjective). Therefore $L_{\Theta_r^T} \circ L_{\Theta_l} = \mathrm{id}$ and thus $\Theta_r^T \Theta_l = I$.

Putting this representation of Π in (6.27)–(6.29), multiplying (6.27) and (6.29) from the left with Θ_r^T and introducing $\bar{\mathbf{v}}_1(t) := \Theta_l^T \mathbf{v}_1(t) \in \mathbb{R}^{n-m}$, we obtain the following system

$$
\Theta_r^T M \Theta_r \frac{d}{dt} \bar{\mathbf{v}}_1(t) - \Theta_r^T A \Theta_r \bar{\mathbf{v}}_1(t) = \Theta_r^T(\mathbf{g}_1(t) - AM^{-1}B^T(BM^{-1}B^T)^{-1}\mathbf{g}_2(t)) \tag{6.30}
$$
$$
\Theta_r^T M \Theta_r \bar{\mathbf{v}}_1(0) = \Theta_r^T \mathbf{v}_0, \tag{6.31}
$$

where $\Theta_r^T M \Theta_r$ is nonsingular since Θ_r is injective. Therefore the Picard–Lindelöf existence theorem guarantees that this system has a unique solution.

We can compute a solution $\mathbf{v}_1$ of (6.27)–(6.29) from the solution $\bar{\mathbf{v}}_1(t)$ of (6.30)–(6.31) by multiplying with Θ_r, because $\Theta_r \bar{\mathbf{v}}_1(t) = \Theta_r \Theta_l^T \mathbf{v}_1(t) = \Pi^T \mathbf{v}_1(t) = \mathbf{v}_1(t)$.

The system (6.30)–(6.31) does not contain an algebraic condition: One can understand $L_{\Theta_l^T}$ in an algebraic way as a linear isomorphism from $\ker(B)$ to $\mathbb{R}^{n-m}$. To make this clear, it is sufficient to show that this mapping is surjective, because $\ker(B)$ has the same dimension as $\mathbb{R}^{n-m}$, due to $\dim \ker(B) = n - \dim \mathrm{im}(B) = n - m$. For the proof of surjectivity, let $\mathbf{w} \in \mathbb{R}^{n-m}$ be arbitrarily given. Then it holds $\mathbf{w} = \Theta_l^T(\Theta_r \mathbf{w})$, and the vector $\Theta_r \mathbf{w} = \Pi^T \Theta_r \mathbf{w}$ lies in $\mathrm{im}(\Pi^T) = \ker(B)$.

The algebraic condition $\Pi^T \mathbf{v}_1(t) = \mathbf{v}_1(t)$ of system (6.27)–(6.29) is equivalent to the condition $\mathbf{v}_1(t) \in \ker(B)$. Applying the isomorphism $L_{\Theta_l^T} : \ker(B) \to \mathbb{R}^{n \times m}$ respectively its inverse L_{Θ_r}, we can equivalently look for a solution which takes values in $\mathbb{R}^{n-m}$, thereby eliminating the algebraic condition.

Finally, we prove the estimates (6.15) and (6.16). Considering equation (6.22), it is easy to see that

$$\|\mathbf{v}_2\|_{L^2} \leq c \|\mathbf{g}_2\|_{L^2}, \tag{6.32}$$

for some constant $c > 0 \in \mathbb{R}$. Next, we want to apply Lemma 3.5 to (6.30)–(6.31). To this end, we have to verify that there is some positive constant $\bar{\alpha}$ such that $-\Theta_r^T A \Theta_r - \bar{\alpha} \Theta_r^T M \Theta_r$ is positive semidefinite. Let $\lambda_{\max}(M)$ denote the largest eigenvalue of M; then $\mathbf{v}^T M \mathbf{v} \leq \lambda_{\max}(M) \|\mathbf{v}\|^2$ for all $\mathbf{v} \in \mathbb{R}^n$. Then note that for an arbitrary vector $\mathbf{w} \in \mathbb{R}^{n-m}$ we have seen that the product $\Theta_r \mathbf{w}$ is an element of $\ker(B)$. Thus, by the assumptions on A, the estimate

$$-\mathbf{w}^T \Theta_r^T A \Theta_r \mathbf{w} \geq \alpha \|\Theta_r \mathbf{w}\|^2 \geq \frac{\alpha}{\lambda_{\max}(M)} \mathbf{w}^T \Theta_r^T M \Theta_r \mathbf{w} = \bar{\alpha} \mathbf{w}^T \Theta_r^T M \Theta_r \mathbf{w}$$

follows. Now we can apply Lemma (3.5) and get

$$\begin{aligned}
\|\bar{\mathbf{v}}_1\|_{L^2} \leq {} & \frac{\sqrt{2}}{\sqrt{\bar{\alpha}}} \|(\Theta_r^T M \Theta_r)^{-\frac{1}{2}}\| \|(\Theta_r^T M \Theta_r)^{\frac{1}{2}}\| \|\Theta_r^T \mathbf{v}_0\| \\
& + \frac{2}{\bar{\alpha}} \|(\Theta_r^T M \Theta_r)^{-1}\| (c_1 \|\mathbf{g}_1\|_{L^2} + c_2 \|\mathbf{g}_2\|_{L^2}),
\end{aligned} \tag{6.33}$$

for some constants $c_1, c_2 > 0 \in \mathbb{R}$. Putting (6.32) and (6.33) together, the estimate (6.15) follows easily.

The second estimate (6.16) follows then easily, combining (6.26) and (6.15). $\qquad\square$

Note that the constraint condition (6.14) is fulfilled if the initial condition is taken of the form

$$\begin{pmatrix} \mathbf{v}(0) \\ \mathbf{p}(0) \end{pmatrix} := - \begin{pmatrix} A & B^T \\ B & 0 \end{pmatrix}^{-1} \begin{pmatrix} \mathbf{g}_1(0) \\ \mathbf{g}_2(0) \end{pmatrix}.$$

This is the reason why the stationary solution makes a canonical choice as an initial condition. Using Theorem 6.4 we are now able to prove the following theorem.

Theorem 6.5. *Under the same assumptions as in Theorem 6.4, the solution operator* $\mathcal{S}_1 : C^1(0, T; \mathbb{R}^4) \to H^1(0, T; \mathbb{R}^{n_v})$ *for the Stokes–Brinkman equation (6.7)–(6.8) is well-defined and continuous.*

Proof. The existence of a unique solution of the Stokes–Brinkman equation (6.7)–(6.8) immediately follows from Theorem 6.4. Then $\mathcal{S}_1$ is well-defined and we have the estimate

$$\|\mathbf{v}\|_{L^2} \leq c_1 \|\mathbf{v}_0\| + c_2 \|\mathbf{w}\|_{L^2} \tag{6.34}$$

with some constants $c_1, c_2 > 0$.

Next, we show the continuity of $\mathcal{S}_1$. Let $(\mathbf{w}_k)_{k \in \mathbb{N}}$ denote a sequence in $C^1(0, T; \mathbb{R}^4)$ with $\lim_{k \to \infty} \mathbf{w}_k = \bar{\mathbf{w}}$. We verify that

$$\lim_{k \to \infty} S_1(\mathbf{w}_k) = S_1(\bar{\mathbf{w}}),$$

in $H^1(0, T; \mathbb{R}^{n_v})$.

Let $[\mathbf{v}_k^T(t), \mathbf{p}_k^T(t)]^T$ and $[\bar{\mathbf{v}}^T(t), \bar{\mathbf{p}}^T(t)]^T$ be the solutions of the Stokes–Brinkman equation (6.7)–(6.8) with the inputs $\mathbf{w}_k$ and $\bar{\mathbf{w}}$, respectively. Then $S_1(\mathbf{w}_k) - S_1(\bar{\mathbf{w}}) = \mathbf{v}_k - \bar{\mathbf{v}}$ solves the DAE

$$E\frac{\mathrm{d}}{\mathrm{d}t}\mathbf{x}_k(t) = S\mathbf{x}_k(t) + F(\mathbf{w}_k(t) - \bar{\mathbf{w}}(t)), \tag{6.35}$$

$$(\mathbf{v}_k - \bar{\mathbf{v}})(0) = -P_\mathbf{v}(S^{-1}F(\mathbf{w}_k(0) - \bar{\mathbf{w}}(0))), \tag{6.36}$$

where $\mathbf{x}_k(t) = [(\mathbf{v}_k - \bar{\mathbf{v}})^T(t), (\mathbf{p}_k - \bar{\mathbf{p}})^T(t)]^T$.

The DAE (6.35)–(6.36) has the form of a Stokes–Brinkman equation and the consistency condition for the initial condition is fulfilled due to the linearity of the system. That means, we can apply Theorem 6.4 to this system and thus can make use of the estimate (6.34). This leads to

$$\|\mathbf{v}_k - \bar{\mathbf{v}}\|_{L^2} \leq c_1\|\mathbf{w}_k(0) - \bar{\mathbf{w}}(0)\| + c_2\|\mathbf{w}_k - \bar{\mathbf{w}}\|_{L^2} \tag{6.37}$$

$$\leq c_1\|\mathbf{w}_k(0) - \bar{\mathbf{w}}(0)\| + c_2\|\mathbf{w}_k - \bar{\mathbf{w}}\|_{H^1} \xrightarrow{k \to \infty} 0 \tag{6.38}$$

which results in convergence in the L^2-norm. From equation (6.35) we have

$$M\frac{\mathrm{d}}{\mathrm{d}t}(\mathbf{v}_k(t) - \bar{\mathbf{v}}(t)) - A(\mathbf{v}_k(t) - \bar{\mathbf{v}}(t)) - B^T(\mathbf{p}_k(t) - \bar{\mathbf{p}}(t)) = \bar{F}_1(\mathbf{w}_k(t) - \bar{\mathbf{w}}(t)).$$

Since M is symmetric and positive definite, we derive the estimate

$$\alpha\|\frac{\mathrm{d}}{\mathrm{d}t}(\mathbf{v}_k(t) - \bar{\mathbf{v}}(t))\| \leq \|A\|\|\mathbf{v}_k(t) - \bar{\mathbf{v}}(t)\| + \|B^T\|\|\mathbf{p}_k(t) - \bar{\mathbf{p}}(t)\| + \|\bar{F}_1\|\|\mathbf{w}_k(t) - \bar{\mathbf{w}}(t)\|, \tag{6.39}$$

where $\alpha = \lambda_{\min}(M) > 0$ is the smallest eigenvalue of M. Using Young's inequality we obtain

$$\|\frac{\mathrm{d}}{\mathrm{d}t}(\mathbf{v}_k(t) - \bar{\mathbf{v}}(t))\|^2 \leq c\left(\|\mathbf{v}_k(t) - \bar{\mathbf{v}}(t)\|_?^2 + \|\mathbf{p}_k(t) - \bar{\mathbf{p}}(t)\|^2 + \|\mathbf{w}_k(t) - \bar{\mathbf{w}}(t)\|^2\right),$$

with $c = \frac{1}{\alpha^2}\max\{4\|A\|_2^2, 4\|B\|_2^2, 2\|\bar{F}_1\|_2^2\}$.

Integration over $(0, T)$ leads to

$$\|\frac{\mathrm{d}}{\mathrm{d}t}(\mathbf{v}_k - \bar{\mathbf{v}})\|_{L_2}^2 \leq c\left(\|\mathbf{v}_k - \bar{\mathbf{v}}\|_{L_2}^2 + \|\mathbf{p}_k - \bar{\mathbf{p}}\|_{L_2}^2 + \|\mathbf{w}_k - \bar{\mathbf{w}}\|_{L_2}^2\right). \tag{6.40}$$

Finally, we have to deal with the right-hand side of (6.40). From (6.38) and the L^2-convergence of $(\mathbf{w}_k)$ it yields

$$\|\mathbf{v}_k - \bar{\mathbf{v}}\|_{L_2}^2 \overset{k\to\infty}{\longrightarrow} 0,$$

$$\|\mathbf{w}_k - \bar{\mathbf{w}}_k\|_{L_2}^2 \overset{k\to\infty}{\longrightarrow} 0.$$

For the remaining term $\|\mathbf{p}_k - \bar{\mathbf{p}}\|_{L_2}$ we use the estimate (6.16) of Theorem 6.4 which leads to

$$\|\mathbf{p}_k - \bar{\mathbf{p}}\|_{L_2} \leq c_1 \|\mathbf{w}_k(0) - \bar{\mathbf{w}}(0)\| + c_4 \|\mathbf{w}_k - \bar{\mathbf{w}}\|_{L_2} + c_5 \|\frac{\mathrm{d}}{\mathrm{d}t}(\mathbf{w}_k - \bar{\mathbf{w}})\|_{L_2} \overset{k\to\infty}{\longrightarrow} 0. \qquad (6.41)$$

Thus, $\lim_{k\to\infty} \|\mathbf{p}_k - \bar{\mathbf{p}}\|_{L_2} = 0$, which completes the proof. $\qquad\square$

6.2 Solvability of the semi-discretized advection-diffusion equation

Next, we study the solution operator $\mathcal{S}_2$ for the advection-diffusion equation (6.9)–(6.10). First, we present some lemmas which are required for later proofs.

Lemma 6.6. *Let* $(\mathbf{v}_k)_{k\in\mathbb{N}} \subset C^0(0,T;\mathbb{R}^n)$ *be a given sequence and* $\bar{\mathbf{v}} \in C^0(0,T;\mathbb{R}^n)$ *with* $\lim_{k\to\infty} \mathbf{v}_k = \bar{\mathbf{v}}$ *in* $C^0(0,T;\mathbb{R}^n)$. *Furthermore, let* $\mathcal{A} : \mathbb{R}^n \to \mathbb{R}^{m\times m}$ *be continuously differentiable. Then it holds*

$$\|\mathcal{A} \circ \mathbf{v}_k - \mathcal{A} \circ \bar{\mathbf{v}}\|_{C^0} \overset{k\to\infty}{\longrightarrow} 0. \qquad (6.42)$$

Proof. The function $\bar{\mathbf{v}}$ is continuous on the interval $[0,T]$ which implies that the image of $\bar{\mathbf{v}}$ is compact. Therefore, we can assume for the image D, that $D \subset \bar{B}_R(0) \subset \mathbb{R}^n$ for an appropriately chosen $R > 0$. Here, $\bar{B}_R(0)$ denotes the closed ball around zero with the radius R, i.e.

$$\bar{B}_R(0) = \{\mathbf{x} \in \mathbb{R}^n \mid \|\mathbf{x}\| \leq R\}.$$

The requirement $\|\mathbf{v}_k - \bar{\mathbf{v}}\|_{C^0} \to 0$ implies that there exists $k_0 > 0$ such that

$$\|\mathbf{v}_k - \bar{\mathbf{v}}\|_{C^0} \leq 1$$

for all $k \geq k_0$. Then it follows, that the images of $\bar{\mathbf{v}}$ and of the maps $\mathbf{v}_k$ for $k \geq k_0$ are included in the compact ball $\bar{B}_{R+1}(0)$.

Since $\mathcal{A}$ was supposed to be continuously differentiable, the derivative of $\mathcal{A}$ admits a maximum L on $\bar{B}_{R+1}(0)$ and $\mathcal{A}$ is Lipschitz on $\bar{B}_{R+1}(0)$ with the Lipschitz constant L.

Finally, for $k \geq k_0$ we obtain

$$\|\mathcal{A} \circ \bar{\mathbf{v}} - \mathcal{A} \circ \mathbf{v}_k\|_{C^0} = \sup_{t\in[0,T]} \|\mathcal{A}(\bar{\mathbf{v}}(t)) - \mathcal{A}(\mathbf{v}_k(t))\|_2 \leq L\|\mathbf{v}_k - \bar{\mathbf{v}}\|_{C^0} \overset{k\to\infty}{\longrightarrow} 0. \qquad \square$$

Corollary 6.7. *Let* $(\mathbf{v}_k)_{k\in\mathbb{N}} \subset C^0(0,T;\mathbb{R}^n)$ *be a given sequence and* $\bar{\mathbf{v}} \in C^0(0,T;\mathbb{R}^n)$ *with* $\lim_{k\to\infty}\mathbf{v}_k = \bar{\mathbf{v}}$ *in* $C^0(0,T;\mathbb{R}^n)$. *Furthermore, let* $\mathcal{A} : \mathbb{R}^n \to \mathbb{R}^{m\times m}$ *be continuously differentiable. Then it holds*

$$\|\mathcal{A}\circ\mathbf{v}_k - \mathcal{A}\circ\bar{\mathbf{v}}\|_{L^2} \overset{k\to\infty}{\longrightarrow} 0.$$

Proof. The convergence follows from $\|\mathcal{A}\circ\mathbf{v}_k - \mathcal{A}\circ\bar{\mathbf{v}}\|_{L^2} \leq C\|\mathcal{A}\circ\mathbf{v}_k - \mathcal{A}\circ\bar{\mathbf{v}}\|_{C^0}$ and Lemma 6.6. $\qquad\square$

Lemma 6.8. *Let* $(\mathbf{v}_k)_{k\in\mathbb{N}} \subset C^0(0,T;\mathbb{R}^n)$ *be a given sequence and* $\bar{\mathbf{v}} \in C^0(0,T;\mathbb{R}^n)$ *with* $\lim_{k\to\infty}\mathbf{v}_k = \bar{\mathbf{v}}$ *in* $C^0(0,T;\mathbb{R}^n)$. *Furthermore, let* $\mathcal{A} : \mathbb{R}^n \to \mathbb{R}^{m\times m}$ *be continuously differentiable. Then it holds*

$$\int_0^t \|\mathcal{A}(\mathbf{v}_k(s))\|_2\,\mathrm{d}s \overset{k\to\infty}{\longrightarrow} \int_0^t \|\mathcal{A}(\bar{\mathbf{v}}(s))\|_2\,\mathrm{d}s. \tag{6.43}$$

Proof. It follows from Lemma 6.6 that

$$\|\|\mathcal{A}\circ\mathbf{v}_k\|_2 - \|\mathcal{A}\circ\bar{\mathbf{v}}\|_2\|_{C^0} = \sup_{t\in[0,T]}\big|\,\|\mathcal{A}(\mathbf{v}_k(t))\|_2 - \|\mathcal{A}(\bar{\mathbf{v}}(t))\|_2\,\big| \leq \sup_{t\in[0,T]}\|\mathcal{A}(\mathbf{v}_k(t)) - \mathcal{A}(\bar{\mathbf{v}}(t))\|_2$$

$$\leq \|\mathcal{A}\circ\mathbf{v}_k - \mathcal{A}\circ\bar{\mathbf{v}}\|_{C^0} \overset{k\to\infty}{\longrightarrow} 0. \tag{6.44}$$

This implies that the sequence of integrands $\|\mathcal{A}(\mathbf{v}_k(s))\|_2$ converges uniformly and therefore also the integral expression in (6.43) converges. $\qquad\square$

Finally, we establish the properties of the solution operator $\mathcal{S}_2$.

Theorem 6.9. *Assume that* $A : \mathbb{R}^{n_v} \to \mathbb{R}^{n_c\times n_c}$ *is twice continuously Fréchet differentiable. Then the solution operator* $\mathcal{S}_2 : C^0(0,T;\mathbb{R}^{n_v}) \to C^1(0,T;\mathbb{R}^{n_c})$ *is well-defined and continuous.*

Proof. Trivially, we can estimate

$$\|A(\mathbf{v}(t))\mathbf{y}(t) - A(\mathbf{v}(t))\mathbf{x}(t)\| \leq \|A(\mathbf{v}(t))\|_2\|\mathbf{y}(t) - \mathbf{x}(t)\|. \tag{6.45}$$

Since $A \circ \mathbf{v}$ is continuous as a composition of continuous functions, it admits a maximum on the interval $[0,T]$. Then it follows from (6.45), that the right-hand side of (6.9) is Lipschitz continuous. An application of the theorem of Picard–Lindelöf guarantees the well-definedness of the solution operator $\mathcal{S}_2$.

To show that $\mathcal{S}_2$ is continuous, let a sequence $(\mathbf{v}_k)_{k\in\mathbb{N}} \subset C^0(0,T;\mathbb{R}^{n_v})$ with $\lim_{k\to\infty}\mathbf{v}_k = \bar{\mathbf{v}}$ in $C^0(0,T;\mathbb{R}^n_v)$ be given. Let $\mathbf{y}_k(t)$ and $\bar{\mathbf{y}}(t)$ be the solutions of the advection-diffusion equation (6.9)–(6.10) with $\mathbf{v}(t) = \mathbf{v}_k(t)$ and $\mathbf{v}(t) = \bar{\mathbf{v}}(t)$, respectively. Then the difference $\mathbf{y}_k(t) - \bar{\mathbf{y}}(t)$ solves the differential equation

$$\frac{\mathrm{d}}{\mathrm{d}t}(\mathbf{y}_k(t) - \bar{\mathbf{y}}(t)) = A(\mathbf{v}_k(t))\mathbf{y}_k(t) - A(\bar{\mathbf{v}}(t))\bar{\mathbf{y}}(t), \tag{6.46}$$

$$(\mathbf{y}_k(0) - \bar{\mathbf{y}}(0)) = 0. \tag{6.47}$$

Using equation (6.46) we obtain the estimate

$$\|\frac{\mathrm{d}}{\mathrm{d}t}(\mathbf{y}_k(t) - \bar{\mathbf{y}}(t))\| = \|A(\mathbf{v}_k(t))\mathbf{y}_k(t) - A(\bar{\mathbf{v}}(t))\bar{\mathbf{y}}(t) + A(\mathbf{v}_k(t))\bar{\mathbf{y}}(t) - A(\mathbf{v}_k(t))\bar{\mathbf{y}}(t)\|$$
$$\leq \|A(\mathbf{v}_k(t))\|_2\|\mathbf{y}_k(t) - \bar{\mathbf{y}}(t)\| + \|A(\mathbf{v}_k(t)) - A(\bar{\mathbf{v}}(t))\|_2\|\bar{\mathbf{y}}(t)\|.$$
$$(6.48)$$

Now we integrate over $(0, t)$ and end up with

$$\|\mathbf{y}_k(t) - \bar{\mathbf{y}}(t)\| \leq \int_0^t \|A(\mathbf{v}_k(s))\|_2\|\mathbf{y}_k(s) - \bar{\mathbf{y}}(s)\|\,\mathrm{d}s + \int_0^t \|A(\mathbf{v}_k(s)) - A(\bar{\mathbf{v}}(s))\|_2\|\bar{\mathbf{y}}(s)\|\,\mathrm{d}s.$$

Gronwall's Lemma implies the estimate

$$\|\mathbf{y}_k(t) - \bar{\mathbf{y}}(t)\| \leq \int_0^t \|A(\mathbf{v}_k(s)) - A(\bar{\mathbf{v}}(s))\|_2\|\bar{\mathbf{y}}(s)\|\,\mathrm{d}s \exp(\int_0^t \|A(\mathbf{v}_k(s))\|_2\,\mathrm{d}s).$$

Furthermore, using the Hölder inequality we obtain

$$\|\mathbf{y}_k(t) - \bar{\mathbf{y}}(t)\| \leq \|A(\mathbf{v}_k) - A(\bar{\mathbf{v}})\|_{L^2}\|\bar{\mathbf{y}}\|_{L^2} \exp(\int_0^t \|A(\mathbf{v}_k(s))\|_2\,\mathrm{d}s).$$

Then it follows from Corollary 6.7 and Lemma 6.8 that

$$\|\mathbf{y}_k - \bar{\mathbf{y}}\|_{C^0} \xrightarrow{k\to\infty} 0 \qquad (6.49)$$

which shows convergence in the C^0-norm.

For the convergence of the derivative we consider the estimate (6.48) again. Using (6.42), (6.44) and (6.49) also the convergence for the derivatives in the C^0-norm follows. This completes the proof. □

6.3 The existence of a minimizer

In this section we finally establish the main result about the functional (6.11). At first, we start with a useful lemma. This well-known fact can for instance be found in [62].

Lemma 6.10. *Let B_1, B_2 be Banach spaces and let $\mathcal{T} : B_1 \to B_2$ be a linear mapping. Then $\mathcal{T}$ is norm continuous if and only if $\mathcal{T}$ is weakly continuous.*

Proof. Let $\mathcal{T}$ be norm continuous and let $f \in B_2^*$ be arbitrarily chosen, where B_2^* denotes the topological dual space inlcuding all linear, continuous functionals. Then the mapping $f \circ \mathcal{T} \in B_1^*$. Indeed, $f \circ \mathcal{T} : B_1 \to \mathbb{R}$ is obviously linear and continuous as a composition of the linear and continuous mappings $\mathcal{T}$ and f. Recall now the definition of weak convergence: A sequence $(x_k)_{k\in\mathbb{N}} \subset B_1$ converging weakly to x is equivalent to the strong convergence $\varphi(x_k) \xrightarrow{k\to\infty} \varphi(x)$ for all $\varphi \in B_1^*$. Then for a sequence $(x_k)_{k\in\mathbb{N}} \subset B_1$ with $x_k \rightharpoonup x$, we have

$f(\mathcal{T}(x_k)) \overset{k \to \infty}{\longrightarrow} f(\mathcal{T}(x))$, and, hence, $\mathcal{T}(x_k) \rightharpoonup \mathcal{T}(x)$ since f was arbitrarily chosen. Thus, $\mathcal{T}$ is weakly continuous.

Assume now that $\mathcal{T}$ is weakly continuous. Then the graph $G \subset B_1 \times B_2$ of $\mathcal{T}$ is weakly closed. We recall the definition of G being weakly closed: For $b_k \rightharpoonup b$ in B_1 and $\mathcal{T}(b_k) \rightharpoonup w$ in B_2 the condition $\mathcal{T}(b) = w$ has to be fulfilled. This is obviously the case, because $\mathcal{T}$ is weakly continuous.

Furthermore, G is convex due to the linearity of $\mathcal{T}$. Indeed, for $(b_1, b_2), (c_1, c_2) \in G$, it holds $(\lambda b_1 + (1 - \lambda)c_1, \lambda b_2 + (1 - \lambda)c_2) \in G$ due to

$$\mathcal{T}(\lambda b_1 + (1 - \lambda)c_1) = \lambda \mathcal{T}(b_1) + (1 - \lambda)\mathcal{T}(c_1) = \lambda b_2 + (1 - \lambda)c_2.$$

For convex sets, it holds that a weakly closed set is also strongly closed and vice versa. Hence, G is strongly closed. Using Theorem 3.4, the continuity of $\mathcal{T}$ follows immediately. □

Now we present the main result.

Theorem 6.11. *The functional* (6.11) *admits a global minimizer.*

Proof. We use the direct method in the calculus of variations [21], so at first we have to show that the functional is bounded from below. Clearly, in our case $J(\mathbf{u}) \geq 0$ for all $\mathbf{u} \in H^3(0, T; \mathbb{R}^2)$. Therefore there exists a minimizing sequence $(\mathbf{u}_k)_{k \in \mathbb{N}} \subset H^3(0, T; \mathbb{R}^2)$ with

$$\lim_{k \to \infty} J(\mathbf{u}_k) = \inf_{\mathbf{u} \in H^3} J(\mathbf{u}).$$

Due to $J(\mathbf{u}_k) \geq \frac{\sigma}{2}\|\mathbf{u}_k\|_{H^3}^2$, the sequence $(\mathbf{u}_k)_k$ is bounded. This ensures the existence of a weakly convergent subsequence [62]. For simplicity of notation we denote this subsequence also by $(\mathbf{u}_k)$ and write $\lim_{k \to \infty} \mathbf{u}_k = \bar{\mathbf{u}}$.

Since $\mathcal{J}$ is linear and continuous, by Lemma 6.10 it is also weakly continuous and therefore $\mathcal{J}(\mathbf{u}_k)$ converges weakly in $H^2(0, T; \mathbb{R}^4)$. Since the embedding operator $\mathcal{I} : H^2(0, T; \mathbb{R}^4) \to C^1(0, T; \mathbb{R}^4)$ is compact, a subsequence of $\mathcal{I}(\mathcal{J}(\mathbf{u}_k))$ converges strongly to $\mathcal{I}(\mathcal{J}(\bar{\mathbf{u}}))$. Recall that the functional reads

$$J(\mathbf{u}) = J_1(\mathbf{u}) + J_2(\mathbf{u}) = \frac{1}{2}\|\mathcal{S}_2(\mathcal{L}(\mathcal{S}_1(\mathcal{I}(\mathcal{J}(\mathbf{u})))))(T) - \mathbf{c}^{\text{foc}}\|_{M_1}^2 + \frac{\sigma}{2}\|\mathbf{u}\|_{H^3}^2.$$

The solution operators $\mathcal{S}_1, \mathcal{S}_2$ are continuous as well as the compact operator $\mathcal{L} : H^1(0, T; \mathbb{R}^{n_v}) \to C^0(0, T; \mathbb{R}^{n_v})$ and we obtain the convergence

$$\lim_{k \to \infty} J_1(\mathbf{u}_k) = J_1(\bar{\mathbf{u}}).$$

Note, that the norm $\|\cdot\|_{H^3}$ is lower semicontinuous such that it yields

$$\inf_{\mathbf{u} \in H^3} J(\mathbf{u}) = \lim_{k \to \infty} J(\mathbf{u}_k) \geq J(\bar{\mathbf{u}}) \geq \inf_{\mathbf{u} \in H^3} J(\mathbf{u}). \tag{6.50}$$

Thus, from (6.50) we obtain $J(\bar{\mathbf{u}}) = \inf_{\mathbf{u} \in H^3} J(\mathbf{u})$. This completes the proof. □

In numerical experiments one will not work with H^3-controls. Therefore we work with a finite dimensional vector space Υ of cubic splines $s \in C^2(0,T;\mathbb{R}^2)$ with $s(0) = \mathbf{0}$ (with respect to an arbitrary mesh on $[0,T]$). The boundary condition is necessary to simplify the description of the gradient of J; it could also be dispensed with, in exchange for an additional term (see Remark 7.12 below). We equip Υ with the L^2 scalar product. Note that, as a finite dimensional space, it is complete and therefore a Hilbert space with respect to this scalar product.

Using Υ as a control space, we can prove the following theorem.

Theorem 6.12. *The functional*

$$J(\mathbf{u}) = J_1(\mathbf{u}) + J_2(\mathbf{u}) = \frac{1}{2}\|\mathcal{S}_2(\mathcal{L}(\mathcal{S}_1(\mathcal{J}(\mathbf{u}))))(T) - \mathbf{c}^{foc}\|_{M_1}^2 + \frac{\sigma}{2}\|\mathbf{u}\|_{L^2}^2 \qquad (6.51)$$

where $\mathcal{J}$ is restricted to Υ, admits a global minimizer.

Again, $\mathcal{S}_1 : C^1(0,T;\mathbb{R}^4) \to H^1(0,T;\mathbb{R}^{n_v})$ and $\mathcal{S}_2 : C^0(0,T;\mathbb{R}^{n_p}) \to C^1(0,T;\mathbb{R}^{n_c})$ are the solution operators of the Stokes–Brinkman equation (6.7)–(6.8) and advection-diffusion equation (6.9)–(6.10), respectively.

Proof. Obviously, $\mathcal{J}$ is a linear mapping. It is furthermore continuous. Indeed, for all $\mathbf{u} \in \Upsilon \subset C^2(0,T;\mathbb{R}^2)$ we have

$$\|\mathcal{J}(\mathbf{u})\|_{C^1} = \left\|\begin{pmatrix} \mathbf{u} \\ \frac{d}{dt}\mathbf{u} \end{pmatrix}\right\|_{C^1} = \left\|\begin{pmatrix} \mathbf{u} \\ \frac{d}{dt}\mathbf{u} \end{pmatrix}\right\|_{C^0} + \left\|\begin{pmatrix} \frac{d}{dt}\mathbf{u} \\ \frac{d^2}{dt^2}\mathbf{u} \end{pmatrix}\right\|_{C^0}$$

$$\leq \|\mathbf{u}\|_{C^0} + 2\|\frac{d}{dt}\mathbf{u}\|_{C^0} + \|\frac{d^2}{dt^2}\mathbf{u}\|_{C^0} \leq 2\|\mathbf{u}\|_{C^2}.$$

The space Υ is finite-dimensional which implies that all norms on Υ are equivalent. Hence, there exists a constant $c > 0$ such that

$$\|\mathbf{u}\|_{C^2} \leq c\|\mathbf{u}\|_{L^2}$$

for all $\mathbf{u} \in \Upsilon$. This implies that

$$\|\mathcal{J}(\mathbf{u})\|_{C^1} \leq 2c\|\mathbf{u}\|_{L^2}$$

which shows the continuity of $\mathcal{J}$. The rest of the proof is analogous to the proof of Theorem 6.11 with the only difference that the minimizing sequence $(\mathbf{u}_k)_k \subset \Upsilon$ has a strongly convergent subsequence. This follows from the fact that in a finite dimensional Hilbert space, a weakly convergent sequence converges strongly as well. $\square$

In the same way, the following corollary can be shown.

Corollary 6.13. *Using constant controls $\mathbf{u} \in \mathbb{R}^2$, the functional*

$$J(\mathbf{u}) = \frac{1}{2}\|\mathcal{S}_2(\mathcal{L}(\mathcal{S}_1(\mathcal{G}(\mathbf{u}))))(T) - \mathbf{c}^{foc}\|_{M_1}^2 + \frac{\sigma}{2}\|\mathbf{u}\|_2^2.$$

admits a global minimizer, where the linear, continuous operator $\mathcal{G} : \mathbb{R}^2 \to C^1(0,T;\mathbb{R}^4)$ is defined as $\mathcal{G}(\mathbf{u}) = \left(\begin{smallmatrix} \mathbf{u} \\ \mathbf{0} \end{smallmatrix}\right)$.

6.4 Differentiability of the solution operators

Before we consider the Lagrange framework, we prove that the solution operators $\mathcal{S}_1, \mathcal{S}_2$ from sections 6.1 and 6.2 are twice continuously Fréchet differentiable. This ensures that the functional J is as well twice contiunously Fréchet differentiable.

Theorem 6.14. *The solution operator $\mathcal{S}_1 : C^1(0, T; \mathbb{R}^4) \to H^1(0, T; \mathbb{R}^{n_v})$ of the discretized Stokes–Brinkman equation (6.7)–(6.8) is twice continuously Fréchet differentiable.*

Proof. Consider the operator $\mathcal{S}_1$ for the Stokes–Brinkman equation (6.7)–(6.8). It has been shown in Theorem 6.5 that $\mathcal{S}_1$ is continuous. We start with the first Fréchet derivative of $\mathcal{S}_1$ at $\mathbf{u}$, denoted by $D\mathcal{S}_1(\mathbf{u})$. Let $\mathbf{x_{u+h}} = [\mathbf{v}_{\mathbf{u+h}}^T, \mathbf{p}_{\mathbf{u+h}}^T]^T$ denote the solution of (6.7)–(6.8) corresponding to the input $\mathbf{u} + \mathbf{h} \in C^1(0, T; \mathbb{R}^4)$, and let $\mathbf{x_u} = [\mathbf{v}_{\mathbf{u}}^T, \mathbf{p}_{\mathbf{u}}^T]^T$ be the solution of (6.7)–(6.8) corresponding to the input $\mathbf{u}$, respectively. Then the difference $\mathbf{w_{u,h}} = [(\mathbf{w}_{\mathbf{u,h}}^{(1)})^T, (\mathbf{w}_{\mathbf{u,h}}^{(2)})^T]^T := [(\mathbf{v_{u+h}} - \mathbf{v_u})^T, (\mathbf{p_{u+h}} - \mathbf{p_u})^T]^T$ fulfills the following DAE

$$E\frac{\mathrm{d}}{\mathrm{d}t}\mathbf{w_{u,h}}(t) = S\mathbf{w_{u,h}}(t) + F\mathbf{h}(t), \tag{6.52}$$

$$\mathbf{w}_{\mathbf{u,h}}^{(1)}(0) = -P_{\mathbf{v}}(S^{-1}F\mathbf{h}(0)). \tag{6.53}$$

We verify that $D\mathcal{S}_1(\mathbf{u})\mathbf{h} = \mathbf{w}_{\mathbf{u,h}}^{(1)}$. Considering system (6.52)–(6.53) it follows immediately that $\mathbf{w_{u,h}}$ is linear in $\mathbf{h}$. Furthermore, from the estimate (6.34) it follows that

$$\|\mathbf{w}_{\mathbf{u,h}}^{(1)}\|_{L^2} \le c_1\|\mathbf{h}(0)\| + c_2\|\mathbf{h}\|_{L^2} \le c\|\mathbf{h}\|_{C^1}, \tag{6.54}$$

for some constant $c > 0$ and using the estimate (6.40), the estimate in (6.41) and (6.54), we obtain

$$\|\frac{\mathrm{d}}{\mathrm{d}t}\mathbf{w}_{\mathbf{u,h}}^{(1)}\|_{L^2} \le \tilde{c}(\|\mathbf{w}_{\mathbf{u,h}}^{(1)}\|_{L^2} + \|\mathbf{h}(0)\| + \|\mathbf{h}\|_{L^2} + \|\frac{\mathrm{d}}{\mathrm{d}t}\mathbf{h}\|_{L^2}),$$

$$\le \hat{c}\|\mathbf{h}\|_{C^1},$$

for constants $\tilde{c}, \hat{c} > 0$. Thus, $\mathbf{w}_{\mathbf{u,h}}^{(1)}$ is continuous with respect to $\mathbf{h}$ and we verified $D\mathcal{S}_1(\mathbf{u})\mathbf{h} = \mathbf{w}_{\mathbf{u,h}}^{(1)}$.

The continuity of $D\mathcal{S}_1(\mathbf{u})\mathbf{h}$ with respect to $\mathbf{u}$ follows directly, because equations (6.52), (6.53) for $\mathbf{w_{u,h}}$ do not depend on $\mathbf{u}$ any longer. The computation of the second derivative is therefore easy, because it follows immediately that $D^2\mathcal{S}_1(\mathbf{u})(\mathbf{h}_1, \mathbf{h}_2) = 0$. □

Next, we study the Fréchet differentiability of the solution operator $\mathcal{S}_2$ for the advection-diffusion equation (6.9)–(6.10) studied in section 6.2.

Theorem 6.15. *Let $A : \mathbb{R}^{n_v} \to \mathbb{R}^{n_c \times n_c}$ be twice uniformly differentiable. Then the solution operator $\mathcal{S}_2 : C^0(0, T; \mathbb{R}^{n_v}) \to C^1(0, T; \mathbb{R}^{n_c})$ of the discretized advection-diffusion equation (6.9)–(6.10) is twice continuously Fréchet differentiable.*

Proof. Consider the solution operator $\mathcal{S}_2$ for the advection diffusion equation (6.9)–(6.10) which is continuous by Theorem 6.9.

First Fréchet derivative

For the computation of the first Fréchet derivative of $\mathcal{S}_2$ we consider the difference $\mathbf{w}_{\mathbf{v},\mathbf{h}} := \mathcal{S}_2(\mathbf{v} + \mathbf{h}) - \mathcal{S}_2(\mathbf{v}) = \mathbf{y}_{\mathbf{v}+\mathbf{h}} - \mathbf{y}_{\mathbf{v}}$, where $\mathbf{y}_{\mathbf{v}}$ is the solution of (6.9)–(6.10) and $\mathbf{y}_{\mathbf{v}+\mathbf{h}}$ solves the initial value problem

$$\frac{\mathrm{d}}{\mathrm{d}t}\mathbf{y}_{\mathbf{v}+\mathbf{h}}(t) = A(\mathbf{v}(t) + \mathbf{h}(t))\mathbf{y}_{\mathbf{v}+\mathbf{h}}(t), \tag{6.55}$$

$$\mathbf{y}_{\mathbf{v}+\mathbf{h}}(0) = \mathbf{y}_0. \tag{6.56}$$

Subtracting equations (6.9)–(6.10) from (6.55)–(6.56) leads to the equations

$$\frac{\mathrm{d}}{\mathrm{d}t}\mathbf{w}_{\mathbf{v},\mathbf{h}}(t) = A(\mathbf{v}(t) + \mathbf{h}(t))\mathbf{y}_{\mathbf{v}+\mathbf{h}}(t) - A(\mathbf{v}(t))\mathbf{y}_{\mathbf{v}}(t),$$

$$\mathbf{w}_{\mathbf{v},\mathbf{h}}(0) = \mathbf{0}.$$

We consider the Taylor expansion of $A(\mathbf{v}(t) + \mathbf{h}(t))$ at $\mathbf{v}(t)$ given by

$$A(\mathbf{v}(t) + \mathbf{h}(t)) = A(\mathbf{v}(t)) + DA(\mathbf{v}(t))\mathbf{h}(t) + R_{\mathbf{v}(t)}(\mathbf{h}(t)).$$

Then we obtain

$$\frac{\mathrm{d}}{\mathrm{d}t}\mathbf{w}_{\mathbf{v},\mathbf{h}}(t) = A(\mathbf{v}(t))\mathbf{y}_{\mathbf{v},\mathbf{h}}(t) + (DA(\mathbf{v}(t))\mathbf{h}(t))\mathbf{y}_{\mathbf{v}+\mathbf{h}}(t) + R_{\mathbf{v}(t)}(\mathbf{h}(t))\mathbf{y}_{\mathbf{v}+\mathbf{h}}(t) - A(\mathbf{v}(t))\mathbf{y}_{\mathbf{v}}(t)$$

$$\mathbf{w}_{\mathbf{v},\mathbf{h}}(0) = \mathbf{0}.$$

If we insert $(DA(\mathbf{v}(t))\mathbf{h}(t))\mathbf{y}_{\mathbf{v}}(t) - (DA(\mathbf{v}(t))\mathbf{h}(t))\mathbf{y}_{\mathbf{v}}(t)$, we arrive at

$$\begin{aligned}
\frac{\mathrm{d}}{\mathrm{d}t}\mathbf{w}_{\mathbf{v},\mathbf{h}}(t) &= A(\mathbf{v}(t))\mathbf{w}_{\mathbf{v},\mathbf{h}}(t) + (DA(\mathbf{v}(t))\mathbf{h}(t))\mathbf{y}_{\mathbf{v}}(t) \\
&\quad + (DA(\mathbf{v}(t))\mathbf{h}(t))(\mathbf{y}_{\mathbf{v}+\mathbf{h}}(t) - \mathbf{y}_{\mathbf{v}}(t)) + R_{\mathbf{v}(t)}(\mathbf{h}(t))\mathbf{y}_{\mathbf{v}+\mathbf{h}}(t),
\end{aligned} \tag{6.57}$$

$$\mathbf{w}_{\mathbf{v},\mathbf{h}}(0) = \mathbf{0}.$$

Considering (6.57) it seems natural to conjecture that the derivative $D\mathcal{S}_2(\mathbf{x})\mathbf{h}$ solves the system

$$\frac{\mathrm{d}}{\mathrm{d}t}\mathbf{z}_{\mathbf{v},\mathbf{h}}(t) = A(\mathbf{v}(t))\mathbf{z}_{\mathbf{v},\mathbf{h}}(t) + (DA(\mathbf{v}(t))\mathbf{h}(t))\mathbf{y}_{\mathbf{v}}(t), \tag{6.58}$$

$$\mathbf{z}_{\mathbf{v},\mathbf{h}}(0) = \mathbf{0}, \tag{6.59}$$

which has a unique solution by the existence theorem of Picard–Lindelöf. We will prove this claim by showing the following results: First, we show that the mapping $\mathbf{h} \mapsto \mathbf{z}_{\mathbf{v},\mathbf{h}}$ is linear and continuous. Second, if we define the remainder as $\mathbf{r}_{\mathbf{v},\mathbf{h}} := \mathbf{w}_{\mathbf{v},\mathbf{h}} - \mathbf{z}_{\mathbf{v},\mathbf{h}}$, we show that $\frac{\|\mathbf{r}_{\mathbf{v},\mathbf{h}}\|_{C^1}}{\|\mathbf{h}\|_{C^0}} \to 0$ for $\mathbf{h} \to 0$. We begin with the properties of the mapping $\mathbf{h} \mapsto \mathbf{z}_{\mathbf{v},\mathbf{h}}$.

Considering system (6.58)–(6.59), it is easy to see, that $\mathbf{z}_{\mathbf{v},\mathbf{h}}$ is linear with respect to $\mathbf{h}$. Furthermore, we derive the estimate

$$\begin{aligned}
\left\|\frac{\mathrm{d}}{\mathrm{d}t}\mathbf{z}_{\mathbf{v},\mathbf{h}}(t)\right\| &\leq \|A(\mathbf{v}(t))\|_2\|\mathbf{z}_{\mathbf{v},\mathbf{h}}(t)\| + \|DA(\mathbf{v}(t))\mathbf{h}(t)\|\|\mathbf{y}_{\mathbf{v}}(t)\|, \\
&\leq \|A(\mathbf{v}(t))\|_2\|\mathbf{z}_{\mathbf{v},\mathbf{h}}(t)\| + \|DA(\mathbf{v}(t))\|_{\mathcal{L}(\mathbb{R}^{n_v},\mathbb{R}^{n_c \times n_c})}\|\mathbf{h}(t)\|\|\mathbf{y}_{\mathbf{v}}(t)\|.
\end{aligned} \tag{6.60}$$

Integration over $(0, t)$ leads to

$$\|\mathbf{z_{v,h}}(t)\| \le \int_0^t \|A(\mathbf{v}(s))\|_2 \|\mathbf{z_{v,h}}(s)\| + \|DA(\mathbf{v}(s))\|_{\mathcal{L}(\mathbb{R}^{n_v}, \mathbb{R}^{n_c \times n_c})} \|\mathbf{h}(s)\| \|\mathbf{y_v}(s)\| \, \mathrm{d}s$$

and with an application of Gronwall's inequality we get

$$\|\mathbf{z_{v,h}}(t)\| \le \left(\int_0^T \|DA(\mathbf{v}(s))\|_{\mathcal{L}(\mathbb{R}^{n_v}, \mathbb{R}^{n_c \times n_c})} \|\mathbf{h}(s)\| \|\mathbf{y_v}(s)\| \, \mathrm{d}s \right) \exp \int_0^T \|A(\mathbf{v}(s))\|_2 \, \mathrm{d}s$$

$$\le \|\mathbf{h}\|_{C^0} \left(\int_0^T \|DA(\mathbf{v}(s))\|_{\mathcal{L}(\mathbb{R}^{n_v}, \mathbb{R}^{n_c \times n_c})} \|\mathbf{y_v}(s)\| \, \mathrm{d}s \right) \exp \int_0^T \|A(\mathbf{v}(s))\|_2 \, \mathrm{d}s.$$

Using this result and (6.60), we obtain furthermore

$$\|\frac{\mathrm{d}}{\mathrm{d}t} \mathbf{z_{v,h}}(t)\| \le c \|\mathbf{h}\|_{C^0},$$

for some constant $c > 0$ which only depends on $\mathbf{v}$. Thus, the continuity of $\mathbf{z_{v,h}}$ with respect to $\mathbf{h}$ follows.

Now, we discuss the remainder $\mathbf{r_{v,h}}$. Subtracting system (6.58)–(6.59) from (6.57), we obtain the equations

$$\frac{\mathrm{d}}{\mathrm{d}t} \mathbf{r_{v,h}}(t) = A(\mathbf{v}(t)) \mathbf{r_{v,h}}(t) + (DA(\mathbf{v}(t)) \mathbf{h}(t))(\mathbf{y_{v+h}}(t) - \mathbf{y_v}(t)) + R_{\mathbf{v}(t)}(\mathbf{h}(t)) \mathbf{y_{v+h}}(t),$$

$$\mathbf{r_{v,h}}(0) = \mathbf{0}.$$

From this differential equation, we have the estimate

$$\|\frac{\mathrm{d}}{\mathrm{d}t} \mathbf{r_{v,h}}(t)\| \le \|A(\mathbf{v}(t))\|_2 \|\mathbf{r_{v,h}}(t)\| + \|DA(\mathbf{v}(t)) \mathbf{h}(t)\|_2 \|\mathbf{y_{v+h}}(t) - \mathbf{y_v}(t)\|$$
$$+ \|R_{\mathbf{v}(t)}(\mathbf{h}(t))\|_2 \|\mathbf{y_{v+h}}(t)\|. \tag{6.61}$$

Integration over $(0, t)$ leads to

$$\|\mathbf{r_{v,h}}(t)\| \le \int_0^t \|A(\mathbf{v}(s))\|_2 \|\mathbf{r_{v,h}}(s)\| \, \mathrm{d}s + \int_0^t \|DA(\mathbf{v}(s)) \mathbf{h}(s)\|_2 \|\mathbf{y_{v+h}}(s) - \mathbf{y_v}(s)\| \, \mathrm{d}s$$
$$+ \int_0^t \|R_{\mathbf{v}(t)}(\mathbf{h}(t))\|_2 \|\mathbf{y_{v+h}}(s)\| \, \mathrm{d}s.$$

Applying Gronwall's inequality, we get

$$\|\mathbf{r_{v,h}}(t)\| \le \int_0^T \left(\|DA(\mathbf{v}(s)) \mathbf{h}(s)\|_2 \|\mathbf{y_{v+h}}(s) - \mathbf{y_v}(s)\| + \|R_{\mathbf{v}(t)}(\mathbf{h}(t))\|_2 \|\mathbf{y_{v+h}}(s)\| \right) \mathrm{d}s$$
$$\cdot \exp \left(\int_0^T \|A(\mathbf{v}(s))\|_2 \, \mathrm{d}s \right).$$

$$\tag{6.62}$$

We show that the remainder satisfies

$$\frac{\|\mathbf{r}_{\mathbf{v},\mathbf{h}}\|_{C^1}}{\|\mathbf{h}\|_{C^0}} \to 0 \ \text{ for } \ \|\mathbf{h}\|_{C^0} \to 0. \tag{6.63}$$

Considering the first integral in the right-hand side of (6.62), we have

$$\frac{\int_0^T \|DA(\mathbf{v}(s))\mathbf{h}(s)\|_2 \|\mathbf{y}_{\mathbf{v}+\mathbf{h}}(s) - \mathbf{y}_{\mathbf{v}}(s)\| + \|R_{\mathbf{v}(s)}(\mathbf{h}(s))\|_2 \|\mathbf{y}_{\mathbf{v}+\mathbf{h}}(s)\| \, \mathrm{d}s}{\|\mathbf{h}\|_{C^0}}$$

$$\leq \int_0^T \|DA(\mathbf{v}(s))\|_{\mathcal{L}(\mathbb{R}^{n_v}, \mathbb{R}^{n_c \times n_c})} \frac{\|\mathbf{h}(s)\|}{\|\mathbf{h}\|_{C^0}} \|\mathbf{y}_{\mathbf{v}+\mathbf{h}}(s) - \mathbf{y}_{\mathbf{v}}(s)\| + \frac{\|R_{\mathbf{v}(s)}(\mathbf{h}(s))\|_2}{\|\mathbf{h}\|_{C^0}} \|\mathbf{y}_{\mathbf{v}+\mathbf{h}}(s)\| \, \mathrm{d}s$$

$$\leq \int_0^T \|DA(\mathbf{v}(s))\|_{\mathcal{L}(\mathbb{R}^{n_v}, \mathbb{R}^{n_c \times n_c})} \|\mathbf{y}_{\mathbf{v}+\mathbf{h}}(s) - \mathbf{y}_{\mathbf{v}}(s)\| + \frac{\|R_{\mathbf{v}(s)}(\mathbf{h}(s))\|_2}{\|\mathbf{h}\|_{C^0}} \|\mathbf{y}_{\mathbf{v}+\mathbf{h}}(s)\| \, \mathrm{d}s$$

and due to the continuity of $\mathbf{y}_{\mathbf{v}+\mathbf{h}}$ with respect to $\mathbf{h}$ it holds that

$$\int_0^T \|DA(\mathbf{v}(s))\|_{\mathcal{L}(\mathbb{R}^{n_v}, \mathbb{R}^{n_c \times n_c})} \|\mathbf{y}_{\mathbf{v}+\mathbf{h}}(s) - \mathbf{y}_{\mathbf{v}}(s)\| \, \mathrm{d}s \leq c \|\mathbf{y}_{\mathbf{v}+\mathbf{h}} - \mathbf{y}_{\mathbf{v}}\|_{C^0} \to 0 \text{ for } \|\mathbf{h}\|_{C^0} \to 0.$$

For the second term, we derive the estimate

$$\frac{\int_0^T \|R_{\mathbf{v}(s)}(\mathbf{h}(s))\|_2 \|\mathbf{y}_{\mathbf{v}+\mathbf{h}}(s)\| \, \mathrm{d}s}{\|\mathbf{h}\|_{C^0}} \leq \frac{\|R_{\mathbf{v}(\cdot)}(\mathbf{h}(\cdot))\|_{L^2}}{\|\mathbf{h}\|_{C^0}} \|\mathbf{y}_{\mathbf{v}+\mathbf{h}}\|_{L^2}.$$

Due to the uniform differentiability of A and the continuity of $\mathbf{y}_{\mathbf{v}+\mathbf{h}}$ with respect to $\mathbf{h}$ we have

$$\frac{\|R_{\mathbf{v}(\cdot)}(\mathbf{h})\|_{L^2}}{\|\mathbf{h}\|_{C^0}} \to 0, \tag{6.64}$$

$$\|\mathbf{y}_{\mathbf{v}+\mathbf{h}}\|_{L^2} \to \|\mathbf{y}_{\mathbf{v}}\|_{L^2}, \tag{6.65}$$

for $\|\mathbf{h}\|_{C^0} \to 0$. The limit (6.64) follows, because for an arbitrary $\epsilon > 0$, it holds that

$$\frac{\sqrt{\int_0^T \|R_{\mathbf{v}(s)}(\mathbf{h}(s))\|_2^2 \, \mathrm{d}s}}{\sqrt{\|\mathbf{h}\|_{C^0}^2}} = \sqrt{\int_0^T \frac{\|R_{\mathbf{v}(s)}(\mathbf{h}(s))\|_2^2}{\|\mathbf{h}\|_{C^0}^2} \, \mathrm{d}s} \leq \sqrt{\int_0^T \frac{\|R_{\mathbf{v}(s)}(\mathbf{h}(s))\|_2^2}{\|\mathbf{h}(s)\|^2} \, \mathrm{d}s} \leq \epsilon \sqrt{T}.$$

Thus, we have

$$\frac{\|\mathbf{r}_{\mathbf{v},\mathbf{h}}\|_{C^0}}{\|\mathbf{h}\|_{C^0}} \to 0 \ \text{ for } \ \|\mathbf{h}\|_{C^0} \to 0. \tag{6.66}$$

For the derivative, we consider (6.61) again. Then (6.63) can be proved using (6.64), (6.65), (6.66). Thus, $\mathcal{S}_2$ is Fréchet differentiable with $D\mathcal{S}_2(\mathbf{v})\mathbf{h} = \mathbf{z}_{\mathbf{v},\mathbf{h}}$.

Continuity of the first Fréchet derivative

Next we verify the continuity of the Fréchet derivative $DS_2(\mathbf{v})$ with respect to $\mathbf{v}$ by showing $\|DS_2(\mathbf{v}_k) - DS_2(\mathbf{w})\|_{\mathcal{L}(C^0(0,T;\mathbb{R}^{n_v}),C^1(0,T;\mathbb{R}^{n_c}))} \to 0$ for $\mathbf{v}_k \to \mathbf{w}$ in $C^0(0,T;\mathbb{R}^{n_v})$. Let therefore $\mathbf{v}_k \in C^0(0,T;\mathbb{R}^m)$ denote a sequence with $\lim_{k\to\infty} \mathbf{v}_k = \mathbf{w}$ in $C^0(0,T;\mathbb{R}^{n_v})$. We show that, for $\mathbf{h} \in C^0(0,T;\mathbb{R}^{n_v})$ with $\|\mathbf{h}\|_{C^0} \leq 1$, $\|DS_2(\mathbf{v}_k)\mathbf{h} - DS_2(\mathbf{w})\mathbf{h}\|_{C^1}$ is bounded from above by an expression which tends to zero as k goes to infinity and in which $\mathbf{h}$ does not appear. We start by considering the two initial value problems for the input functions $\mathbf{v}_k$ and $\mathbf{w}$ which read

$$\frac{\mathrm{d}}{\mathrm{d}t}\mathbf{z}_{\mathbf{v}_k,\mathbf{h}}(t) = A(\mathbf{v}_k(t))\mathbf{z}_{\mathbf{v}_k,\mathbf{h}}(t) + (DA(\mathbf{v}_k(t))\mathbf{h}(t))\mathbf{y}_{\mathbf{v}_k}(t),$$
$$\mathbf{z}_{\mathbf{v}_k,\mathbf{h}}(0) = \mathbf{0}, \tag{6.67}$$

and

$$\frac{\mathrm{d}}{\mathrm{d}t}\mathbf{z}_{\mathbf{w},\mathbf{h}}(t) = A(\mathbf{w}(t))\mathbf{z}_{\mathbf{w},\mathbf{h}}(t) + (DA(\mathbf{w}(t))\mathbf{h}(t))\mathbf{y}_{\mathbf{w}}(t),$$
$$\mathbf{z}_{\mathbf{w},\mathbf{h}}(0) = \mathbf{0}. \tag{6.68}$$

Subtracting (6.68) from (6.67) gives

$$\frac{\mathrm{d}}{\mathrm{d}t}(\mathbf{z}_{\mathbf{v}_k,\mathbf{h}}(t) - \mathbf{z}_{\mathbf{w},\mathbf{h}}(t)) = A(\mathbf{v}_k(t))\mathbf{z}_{\mathbf{v}_k,\mathbf{h}}(t) - A(\mathbf{w}(t))\mathbf{z}_{\mathbf{w},\mathbf{h}}(t)$$
$$+ A(\mathbf{v}_k(t))\mathbf{z}_{\mathbf{w},\mathbf{h}}(t) - A(\mathbf{v}_k(t))\mathbf{z}_{\mathbf{w},\mathbf{h}}(t)$$
$$+ (DA(\mathbf{v}_k(t))\mathbf{h}(t))\mathbf{y}_{\mathbf{v}_k}(t) - (DA(\mathbf{w}(t))\mathbf{h}(t))\mathbf{y}_{\mathbf{w}}(t)$$
$$+ (DA(\mathbf{v}_k(t))\mathbf{h}(t))\mathbf{y}_{\mathbf{w}}(t) - (DA(\mathbf{v}_k(t))\mathbf{h}(t))\mathbf{y}_{\mathbf{w}}(t),$$
$$(\mathbf{z}_{\mathbf{v}_k,\mathbf{h}} - \mathbf{z}_{\mathbf{w},\mathbf{h}})(0) = \mathbf{0}.$$

Then we derive the estimate

$$\|\frac{\mathrm{d}}{\mathrm{d}t}(\mathbf{z}_{\mathbf{v}_k,\mathbf{h}} - \mathbf{z}_{\mathbf{w},\mathbf{h}})(t)\| \leq \|A(\mathbf{v}_k(t))\|_2\|(\mathbf{z}_{\mathbf{v}_k,\mathbf{h}} - \mathbf{z}_{\mathbf{w},\mathbf{h}})(t)\| + \|A(\mathbf{v}_k(t)) - A(\mathbf{w}(t))\|_2\|\mathbf{z}_{\mathbf{w},\mathbf{h}}(t)\|$$
$$+ \|DA(\mathbf{v}_k(t))\mathbf{h}(t)\|_2\|(\mathbf{y}_{\mathbf{v}_k} - \mathbf{y}_{\mathbf{w}})(t)\|$$
$$+ \|DA(\mathbf{v}_k(t))\mathbf{h}(t) - DA(\mathbf{w}(t))\mathbf{h}(t)\|_2\|\mathbf{y}_{\mathbf{w}}(t)\|. \tag{6.69}$$

Integrating over $(0,t)$ results in

$$\|(\mathbf{z}_{\mathbf{v}_k,\mathbf{h}} - \mathbf{z}_{\mathbf{w},\mathbf{h}})(t)\| \leq \int_0^t \|A(\mathbf{v}_k(s))\|_2\|(\mathbf{z}_{\mathbf{v}_k,\mathbf{h}} - \mathbf{z}_{\mathbf{w},\mathbf{h}})(s)\|\,\mathrm{d}s$$
$$+ \int_0^t \|A(\mathbf{v}_k(s)) - A(\mathbf{w}(s))\|_2\|\mathbf{z}_{\mathbf{w},\mathbf{h}}(s)\|_2\,\mathrm{d}s$$
$$+ \int_0^t \|DA(\mathbf{v}_k(s))\mathbf{h}(s)\|_2\|\mathbf{y}_{\mathbf{v}_k}(s) - \mathbf{y}_{\mathbf{w}}(s)\|\,\mathrm{d}s$$
$$+ \int_0^t \|DA(\mathbf{v}_k(s))\mathbf{h}(s) - DA(\mathbf{w}(s))\mathbf{h}(s)\|_2\|\mathbf{y}_{\mathbf{w}}(s)\|\,\mathrm{d}s.$$

47

Applying Gronwall's Lemma, we obtain

$$\|(\mathbf{z}_{\mathbf{v}_k,\mathbf{h}} - \mathbf{z}_{\mathbf{w},\mathbf{h}})(t)\|$$

$$\leq \left(\int_0^T \|A(\mathbf{v}_k(s)) - A(\mathbf{w}(s))\|_2 \|\mathbf{z}_{\mathbf{w},\mathbf{h}}(s)\| \, ds + \int_0^T \|DA(\mathbf{v}_k(s))\mathbf{h}(s)\|_2 \|\mathbf{y}_{\mathbf{v}_k}(s) - \mathbf{y}_{\mathbf{w}}(s)\| \, ds \right.$$

$$\left. + \int_0^T \|DA(\mathbf{v}_k(s))\mathbf{h}(s) - DA(\mathbf{w}(s))\mathbf{h}(s)\|_2 \|\mathbf{y}_{\mathbf{w}}(s)\| \, ds \right) \exp(\textstyle\int_0^T \|A(\mathbf{v}_k(s))\|_2 \, ds)$$

$$\leq \left(\int_0^T \|A(\mathbf{v}_k(s)) - A(\mathbf{w}(s))\|_2 \, c_{\mathbf{w}} \|\mathbf{h}\|_{C^0} \, ds \right.$$

$$+ \int_0^T \|DA(\mathbf{v}_k(s))\|_{\mathcal{L}(\mathbb{R}^{n_v},\mathbb{R}^{n_c \times n_c})} \|\mathbf{h}\|_{C^0} \|\mathbf{y}_{\mathbf{v}_k}(s) - \mathbf{y}_{\mathbf{w}}(s)\| \, ds$$

$$\left. + \int_0^T \|DA(\mathbf{v}_k(s)) - DA(\mathbf{w}(s))\|_{\mathcal{L}(\mathbb{R}^{n_v},\mathbb{R}^{n_c \times n_c})} \|\mathbf{h}\|_{C^0} \|\mathbf{y}_{\mathbf{w}}(s)\| \, ds \right) \exp(\textstyle\int_0^T \|A(\mathbf{v}_k(s))\|_2 \, ds)$$

$$\leq \left(\int_0^T \|A(\mathbf{v}_k(s)) - A(\mathbf{w}(s))\|_2 \, c_{\mathbf{w}} \, ds + \int_0^T \|DA(\mathbf{v}_k(s))\|_{\mathcal{L}(\mathbb{R}^{n_v},\mathbb{R}^{n_c \times n_c})} \|\mathbf{y}_{\mathbf{v}_k}(s) - \mathbf{y}_{\mathbf{w}}(s)\| \, ds \right.$$

$$\left. + \int_0^T \|DA(\mathbf{v}_k(s)) - DA(\mathbf{w}(s))\|_{\mathcal{L}(\mathbb{R}^{n_v},\mathbb{R}^{n_c \times n_c})} \|\|\mathbf{y}_{\mathbf{w}}(s)\| \, ds \right) \exp(\textstyle\int_0^T \|A(\mathbf{v}_k(s))\|_2 \, ds).$$

This holds true, because we have the estimate $\|\mathbf{z}_{\mathbf{w},\mathbf{h}}\|_{C^1} \leq c_{\mathbf{w}} \|\mathbf{h}\|_{C^0}$ due to the continuity of $\mathbf{z}_{\mathbf{w},\mathbf{h}}$ with respect to $\mathbf{h}$ and because we assumed $\|\mathbf{h}\|_{C^0} \leq 1$.

Using the Hölder inequality we get

$$\|(\mathbf{z}_{\mathbf{v}_k,\mathbf{h}} - \mathbf{z}_{\mathbf{w},\mathbf{h}})(t)\| \leq c_{\mathbf{w}}(T\|A \circ \mathbf{v}_k - A \circ \mathbf{w}\|_{L^2} + \|DA(\mathbf{v}_k)\|_{L^2} \|\mathbf{y}_{\mathbf{v}_k} - \mathbf{y}_{\mathbf{w}}\|_{L^2}$$

$$+ \|DA(\mathbf{v}_k) - DA(\mathbf{w})\|_{L^2} \|\mathbf{y}_{\mathbf{w}}\|_{L^2}) \exp(\int_0^T \|A(\mathbf{v}_k(s))\|_2 \, ds).$$

Next we use some results from section 6.2 to show that $\|\mathbf{z}_{\mathbf{v}_k,\mathbf{h}} - \mathbf{z}_{\mathbf{w},\mathbf{h}}\|_{C^0} \to 0$ for $\mathbf{v}_k \to \mathbf{w}$ in C^0. Corollary 6.7 ensures

$$\|A \circ \mathbf{v}_k - A \circ \mathbf{w}\|_{L^2} \to 0,$$

and Lemma 6.8 implies that

$$\|\mathbf{y}_{\mathbf{v}_k} - \mathbf{y}_{\mathbf{w}}\|_{L^2} \to 0.$$

Furthermore, the limits

$$\|DA(\mathbf{v}_k) - DA(\mathbf{w})\|_{L^2} \to 0,$$
$$\|DA(\mathbf{v}_k)\|_{L^2} \to \|DA(\mathbf{w})\|_{L^2}$$

follow directly from the continuous differentiability of A. Using these results we have proved convergence in the C^0-norm. For the derivative, the estimate (6.69) can be used and we obtain

$$\|\frac{\mathrm{d}}{\mathrm{d}t}(\mathbf{z}_{\mathbf{v}_k,\mathbf{h}} - \mathbf{z}_{\mathbf{w},\mathbf{h}})(t)\| \leq \|A(\mathbf{v}_k(t))\|_2 \|D\mathcal{S}_2(\mathbf{v}_k(t))\mathbf{h}(t) - D\mathcal{S}_2(\mathbf{w}(t))\mathbf{h}(t)\|$$
$$+ \|A(\mathbf{v}_k(t)) - A(\mathbf{w}(t))\|_2\, c_{\mathbf{w}}\|\mathbf{h}\|_{C^0}$$
$$+ \|DA(\mathbf{v}_k(t))\|_{\mathcal{L}(\mathbb{R}^{n_v},\mathbb{R}^{n_c \times n_c})}\|\mathbf{h}\|_{C^0}\|(\mathbf{y}_{\mathbf{v}_k} - \mathbf{y}_{\mathbf{w}})(t)\|$$
$$+ \|DA(\mathbf{v}_k(t)) - DA(\mathbf{w}(t))\|_{\mathcal{L}(\mathbb{R}^{n_v},\mathbb{R}^{n_c \times n_c})}\|\mathbf{h}\|_{C^0}\|\mathbf{y}_{\mathbf{w}}(t)\|$$
$$\leq \|A(\mathbf{v}_k(t))\|_2 \|D\mathcal{S}_2(\mathbf{v}_k(t)) - D\mathcal{S}_2(\mathbf{w}(t))\|_{\mathcal{L}(\mathbb{R}^{n_v},\mathbb{R}^{n_c})}$$
$$+ \|A(\mathbf{v}_k(t)) - A(\mathbf{w}(t))\|_2\, c_{\mathbf{w}}$$
$$+ \|DA(\mathbf{v}_k(t))\|_{\mathcal{L}(\mathbb{R}^{n_v},\mathbb{R}^{n_c \times n_c})}\|(\mathbf{y}_{\mathbf{v}_k} - \mathbf{y}_{\mathbf{w}})(t)\|$$
$$+ \|DA(\mathbf{v}_k(t)) - DA(\mathbf{w}(t))\|_{\mathcal{L}(\mathbb{R}^{n_v},\mathbb{R}^{n_c \times n_c})}\|\mathbf{y}_{\mathbf{w}}(t)\|$$
$$\leq \|A(\mathbf{v}_k)\|_{C^0}\|D\mathcal{S}_2(\mathbf{v}_k) - D\mathcal{S}_2(\mathbf{w})\|_{C^0} + \|A \circ \mathbf{v}_k - A \circ \mathbf{w}\|_{C^0}c_{\mathbf{w}}$$
$$+ \|DA(\mathbf{v}_k)\|_{C^0}\|\mathbf{y}_{\mathbf{v}_k} - \mathbf{y}_{\mathbf{w}}\|_{C^0}$$
$$+ \|DA(\mathbf{v}_k) - DA(\mathbf{w})\|_{C^0}\|\mathbf{y}_{\mathbf{w}}\|_{C^0}$$

and together with

$$\|D\mathcal{S}_2(\mathbf{v}_k) - D\mathcal{S}_2(\mathbf{w})\|_{C^0} \to 0$$

which was shown before, we obtain the convergence of the derivative in the C^0-norm. This proves the continuity of the Fréchet derivative.

Second Fréchet derivative

It still remains to cover the second derivative of $\mathcal{S}_2$. We start with the equations (6.58)–(6.59). Again, we consider the difference $\mathbf{s}_{\mathbf{v},\mathbf{h},\mathbf{w}} := \mathbf{z}_{\mathbf{v}+\mathbf{w},\mathbf{h}} - \mathbf{z}_{\mathbf{v},\mathbf{h}} = D\mathcal{S}_2(\mathbf{v}+\mathbf{w})\mathbf{h} - D\mathcal{S}_2(\mathbf{v})\mathbf{h}$ which fulfills

$$\frac{\mathrm{d}}{\mathrm{d}t}\mathbf{s}_{\mathbf{v},\mathbf{h},\mathbf{w}}(t) = A(\mathbf{v}(t) + \mathbf{w}(t))\mathbf{z}_{\mathbf{v}+\mathbf{w},\mathbf{h}}(t) + (DA(\mathbf{v}(t) + \mathbf{w}(t))\mathbf{h}(t))\mathbf{y}_{\mathbf{v}+\mathbf{w}}(t)$$
$$- A(\mathbf{v}(t))\mathbf{z}_{\mathbf{v},\mathbf{h}}(t) - (DA(\mathbf{v}(t))\mathbf{h}(t))\mathbf{y}_{\mathbf{v}}(t),$$
$$\mathbf{s}_{\mathbf{v},\mathbf{h},\mathbf{w}}(0) = 0.$$

Using the Taylor expansions

$$A(\mathbf{v}(t) + \mathbf{w}(t)) = A(\mathbf{v}(t)) + DA(\mathbf{v}(t))\mathbf{w}(t) + R_{\mathbf{v}(t)}(\mathbf{w}(t)),$$
$$DA(\mathbf{v}(t) + \mathbf{w}(t))\mathbf{h}(t) = DA(\mathbf{v}(t))\mathbf{h}(t) + D^2A(\mathbf{v}(t))(\mathbf{h}(t),\mathbf{w}(t)) + R_{\mathbf{v}(t),\mathbf{h}(t)}(\mathbf{w}(t)),$$

49

we arrive at

$$\frac{d}{dt}\mathbf{s_{v,h,w}}(t) = A(\mathbf{v}(t))\mathbf{s_{v,h,w}}(t) + (DA(\mathbf{v}(t))\mathbf{h}(t))(\mathbf{y_{v+w}} - \mathbf{y_v})(t)$$
$$+ (DA(\mathbf{v}(t))\mathbf{w}(t))(\mathbf{z_{v+w,h}} - \mathbf{z_{v,h}})(t)$$
$$+ D^2A(\mathbf{v}(t))(\mathbf{h}(t), \mathbf{w}(t))\mathbf{y_{v+w}}(t) + R_{\mathbf{v}(t),\mathbf{h}(t)}(\mathbf{w}(t))\mathbf{y_{v+w}}(t)$$
$$+ (DA(\mathbf{v}(t))\mathbf{w}(t))\mathbf{z_{v,h}}(t) + R_{\mathbf{v}(t)}(\mathbf{w}(t))\mathbf{z_{v+w,h}}(t),$$
$$\mathbf{s_{v,h,w}}(0) = 0.$$

From this we conjecture that $D^2S_2(\mathbf{v})(\mathbf{h}, \mathbf{w})$ solves the system

$$\frac{d}{dt}\mathbf{q_{v,h,w}}(t) = A(\mathbf{v}(t))\mathbf{q_{v,h,w}}(t) + (DA(\mathbf{v}(t))\mathbf{h}(t))\mathbf{z_{v,w}}(t) + (DA(\mathbf{v}(t))\mathbf{w}(t))\mathbf{z_{v,h}}(t)$$
$$+ D^2A(\mathbf{v}(t))(\mathbf{h}(t), \mathbf{w}(t))\mathbf{y_v}(t), \tag{6.70}$$
$$\mathbf{q_{v,h,w}}(0) = 0.$$

Then $\boldsymbol{\rho}_{\mathbf{v,h,w}} := \mathbf{s_{v,h,w}} - \mathbf{q_{v,h,w}}$ satisfies the system

$$\frac{d}{dt}\boldsymbol{\rho}_{\mathbf{v,h,w}}(t) = A(\mathbf{v}(t))\boldsymbol{\rho}_{\mathbf{v,h,w}}(t) + (DA(\mathbf{v}(t))\mathbf{h}(t))\mathbf{r_{v,w}}(t)$$
$$+ (DA(\mathbf{v}(t))\mathbf{w}(t))(\mathbf{z_{v+w,h}} - \mathbf{z_{v,h}})(t)$$
$$+ D^2A(\mathbf{v}(t))(\mathbf{h}(t), \mathbf{w}(t))(\mathbf{y_{v+w}} - \mathbf{y_v})(t) + R_{\mathbf{v}(t),\mathbf{h}(t)}(\mathbf{w}(t))\mathbf{y_{v+w}}(t)$$
$$+ R_{\mathbf{v}(t)}(\mathbf{w}(t))\mathbf{z_{v+w,h}}(t),$$
$$\boldsymbol{\rho}_{\mathbf{v,h,w}}(0) = 0. \tag{6.71}$$

It is easy to see that $\mathbf{q_{v,h,w}}$ is linear in $\mathbf{w}$. From (6.70) we derive the estimate

$$\|\frac{d}{dt}\mathbf{q_{v,h,w}}(t)\| \leq \|A(\mathbf{v}(t))\|_2\|\mathbf{q_{v,h,w}}(t)\| + \|DA(\mathbf{v}(t))\mathbf{h}(t)\|_2\|\mathbf{z_{v,w}}(t)\|$$
$$+ \|DA(\mathbf{v}(t))\mathbf{w}(t)\|_2\|\mathbf{z_{v,h}}(t)\| \tag{6.72}$$
$$+ \|D^2A(\mathbf{v}(t))(\mathbf{h}(t), \mathbf{w}(t))\|_2\|\mathbf{y_v}(t)\|.$$

Integration over $(0, t)$ leads to

$$\|\mathbf{q_{v,h,w}}(t)\| \leq \int_0^t \|A(\mathbf{v}(s))\|_2\|\mathbf{q_{v,h,w}}(s)\| + \|DA(\mathbf{v}(s))\mathbf{h}(s)\|_2\|\mathbf{z_{v,w}}(s)\|$$
$$+ \|DA(\mathbf{v}(s))\mathbf{w}(s)\|_2\|\mathbf{z_{v,h}}(s)\| \, ds$$
$$+ \int_0^t \|D^2A(\mathbf{v}(s))(\mathbf{h}(s), \mathbf{w}(s))\|_2\|\mathbf{y_v}(s)\| \, ds,$$

We use Gronwall's Lemma to obtain

$$
\begin{aligned}
\|\mathbf{q}_{\mathbf{v},\mathbf{h},\mathbf{w}}(t)\| \leq &\left(\int_0^T \|DA(\mathbf{v}(s))\mathbf{h}(s)\|_2 \|\mathbf{z}_{\mathbf{v},\mathbf{w}}(s)\| + \|DA(\mathbf{v}(s))\mathbf{w}(s)\|_2 \|\mathbf{z}_{\mathbf{v},\mathbf{h}}(s)\| \right.\\
&\left. + \|D^2 A(\mathbf{v}(s))(\mathbf{h}(s),\mathbf{w}(s))\|_2 \|\mathbf{y}_{\mathbf{v}}(s)\| \, \mathrm{d}s \right) \exp(\int_0^T \|A(\mathbf{v}(s))\|_2 \, \mathrm{d}s)\\
\leq &\exp(\int_0^T \|A(\mathbf{v}(s))\|_2 \, \mathrm{d}s)\Big(\int_0^T \|DA(\mathbf{v}(s))\|_{\mathcal{L}(\mathbb{R}^{n_v},\mathbb{R}^{n_c \times n_c})} \|\mathbf{h}\|_{C^0} c_1 \|\mathbf{w}\|_{C^0}\\
&+ \|DA(\mathbf{v}(s))\|_{\mathcal{L}(\mathbb{R}^{n_v},\mathbb{R}^{n_c \times n_c})} \|\mathbf{w}\|_{C^0} c_2 \|\mathbf{h}\|_{C^0}\\
&+ \|D^2 A(\mathbf{v}(s))\|_{\mathcal{L}(\mathbb{R}^{n_v} \times \mathbb{R}^{n_v},\mathbb{R}^{n_c \times n_c})} \|\mathbf{h}\|_{C^0} \|\mathbf{w}\|_{C^0} \|\mathbf{y}_{\mathbf{v}}(s)\| \, \mathrm{d}s \Big)\\
\leq &\, c_{\mathbf{v}} \|\mathbf{h}\|_{C^0} \|\mathbf{w}\|_{C^0}
\end{aligned}
$$

for constants $c_1, c_2, c_{\mathbf{v}} > 0$. Here, we used the continuity of $\mathbf{z}_{\mathbf{v},\mathbf{w}}$ and $\mathbf{z}_{\mathbf{v},\mathbf{h}}$ with respect to $\mathbf{w}, \mathbf{h}$. Thus, it follows that

$$
\|\mathbf{q}_{\mathbf{v},\mathbf{h},\mathbf{w}}\|_{C^0} \leq c_{\mathbf{v}} \|\mathbf{h}\|_{C^0} \|\mathbf{w}\|_{C^0}. \tag{6.73}
$$

Using (6.73) and the estimate (6.72), we obtain in the same way

$$
\|\frac{\mathrm{d}}{\mathrm{d}t}\mathbf{q}_{\mathbf{v},\mathbf{h},\mathbf{w}}\|_{C^0} \leq c_{\mathbf{v}} \|\mathbf{h}\|_{C^0} \|\mathbf{w}\|_{C^0}. \tag{6.74}
$$

Thus, $\mathbf{q}_{\mathbf{v},\mathbf{h},\mathbf{w}}$ is continuous with respect to $\mathbf{w}$.

Now we discuss the remainder $\boldsymbol{\rho}_{\mathbf{v},\mathbf{h},\mathbf{w}}$. The aim is to find an estimate for the solution of (6.71) using Gronwall's Lemma. We start with

$$
\|\frac{\mathrm{d}}{\mathrm{d}t}\boldsymbol{\rho}_{\mathbf{v},\mathbf{h},\mathbf{w}}(t)\| \leq \|A(\mathbf{v}(t))\|_2 \|\boldsymbol{\rho}_{\mathbf{v},\mathbf{h},\mathbf{w}}(t)\| + \beta(t) \tag{6.75}
$$

where we have defined

$$
\begin{aligned}
\beta(t) := &\|DA(\mathbf{v}(t))\mathbf{h}(t)\|_2 \|\mathbf{r}_{\mathbf{v},\mathbf{w}}(t)\| + \|DA(\mathbf{v}(t))\mathbf{w}(t)\|_2 \|(\mathbf{z}_{\mathbf{v}+\mathbf{w},\mathbf{h}} - \mathbf{z}_{\mathbf{v},\mathbf{h}})(t)\|\\
&+ \|R_{\mathbf{v}(t),\mathbf{h}(t)}(\mathbf{w}(t))\|_2 \|\mathbf{y}_{\mathbf{v}+\mathbf{w}}(t)\| + \|R_{\mathbf{v}(t)}(\mathbf{w}(t))\|_2 \|\mathbf{z}_{\mathbf{v}+\mathbf{w},\mathbf{h}}(t)\|\\
&+ \|D^2 A(\mathbf{v}(t))(\mathbf{h}(t),\mathbf{w}(t))(\mathbf{y}_{\mathbf{v}+\mathbf{w}} - \mathbf{y}_{\mathbf{v}})(t)\|.
\end{aligned}
$$

Integration over $(0,t)$ leads to

$$
\|\boldsymbol{\rho}_{\mathbf{v},\mathbf{h},\mathbf{w}}(t)\| \leq \int_0^t \|A(\mathbf{v}(s))\|_2 \|\boldsymbol{\rho}_{\mathbf{v},\mathbf{h},\mathbf{w}}(s)\| \, \mathrm{d}s + \int_0^t \beta(s) \, \mathrm{d}s,
$$

and an application of Gronwall's Lemma gives the estimate

$$
\|\boldsymbol{\rho}_{\mathbf{v},\mathbf{h},\mathbf{w}}(t)\| \leq \int_0^T \beta(s) \, \mathrm{d}s \exp\left(\int_0^T \|A(\mathbf{v}(s))\|_2 \, \mathrm{d}s \right).
$$

51

Using (6.65), the properties of the remainders and the continuity of $\mathbf{z_{v+w,h}}$ and $\mathbf{y_{v+w}}$, we obtain

$$\frac{\|\mathbf{r_{v,w}}\|_{C^0}}{\|\mathbf{w}\|_{C^0}} \longrightarrow 0, \tag{6.76}$$

$$\frac{\|R_{\mathbf{v}}(\mathbf{w})\|_{C^0}\|\mathbf{z_{v+w,h}}\|_{C^0}}{\|\mathbf{w}\|_{C^0}} \longrightarrow 0, \tag{6.77}$$

$$\frac{\|DA(\mathbf{v})\mathbf{w}\|_{C^0}\|\mathbf{z_{v+w,h}} - \mathbf{z_{v,h}}\|_{C^0}}{\|\mathbf{w}\|_{C^0}} \longrightarrow 0, \tag{6.78}$$

$$\frac{\|R_{\mathbf{v,h}}(\mathbf{w})\|_{C^0}\|\mathbf{y_{v+w}}\|_{C^0}}{\|\mathbf{w}\|_{C^0}} \longrightarrow 0, \tag{6.79}$$

$$\frac{\|D^2A(\mathbf{v})(\mathbf{h},\mathbf{w})\|_{C^0}\|\mathbf{y_{v+w}} - \mathbf{y_v}\|_{C^0}}{\|\mathbf{w}\|_{C^0}} \longrightarrow 0, \tag{6.80}$$

and, therefore,

$$\frac{\|\boldsymbol{\rho_{v,h,w}}\|_{C^0}}{\|\mathbf{w}\|_{C^0}} \longrightarrow 0. \tag{6.81}$$

for $\mathbf{w} \to 0$ in $C^0(0,T;\mathbb{R}^{n_v})$.

It remains to show that also

$$\frac{\|\frac{\mathrm{d}}{\mathrm{d}t}\boldsymbol{\rho_{v,h,w}}\|_{C^0}}{\|\mathbf{w}\|_{C^0}} \longrightarrow 0 \quad \text{for } \|\mathbf{w}\|_{C^0} \longrightarrow 0$$

holds true. Consider the estimate (6.75) again. Using the limits (6.76)–(6.80) and (6.81) we obtain that $\frac{\|\boldsymbol{\rho_{v,h,w}}\|_{C^1}}{\|\mathbf{w}\|_{C^0}} \to 0$ for $\mathbf{w} \to 0$ in $C^0(0,T;\mathbb{R}^{n_v})$.

Continuity of the second Fréchet derivative

Finally, we show also the continuity of $D^2\mathcal{S}_2(\mathbf{v})$ in $\mathbf{v}$. Hence, we consider a sequence $(\mathbf{v}_k)_{k\in\mathbb{N}} \subset C^0(0,T;\mathbb{R}^{n_v})$ with $\mathbf{v}_k \to \mathbf{v}$ for $k \to \infty$, $\mathbf{v} \in C^0(0,T;\mathbb{R}^{n_v})$. The difference $\mathbf{q_{v_k,h,w}} - \mathbf{q_{v,h,w}} = D^2\mathcal{S}_2(\mathbf{v}_k)(\mathbf{h},\mathbf{w}) - D^2\mathcal{S}_2(\mathbf{v})(\mathbf{h},\mathbf{w})$ fulfills

$$\begin{aligned}
&\frac{\mathrm{d}}{\mathrm{d}t}(\mathbf{q_{v_k,h,w}} - \mathbf{q_{v,h,w}})(t) \\
&= A(\mathbf{v}_k(t))\mathbf{q_{v_k,h,w}}(t) - A(\mathbf{v}(t))\mathbf{q_{v,h,w}}(t) + A(\mathbf{v}_k(t))\mathbf{q_{v,h,w}}(t) - A(\mathbf{v}_k(t))\mathbf{q_{v,h,w}}(t) \\
&\quad + DA(\mathbf{v}_k(t))\mathbf{h}(t)\mathbf{z_{v_k,w}}(t) - DA(\mathbf{v}(t))\mathbf{h}(t)\mathbf{z_{v,w}}(t) + DA(\mathbf{v}(t))\mathbf{h}(t)\mathbf{z_{v_k,w}}(t) \\
&\quad - DA(\mathbf{v}(t))\mathbf{h}(t)\mathbf{z_{v_k,w}}(t) + DA(\mathbf{v}_k(t))\mathbf{w}(t)\mathbf{z_{v_k,h}}(t) - DA(\mathbf{v}(t))\mathbf{w}(t)\mathbf{z_{v,h}}(t) \\
&\quad + DA(\mathbf{v}(t))\mathbf{w}(t)\mathbf{z_{v_k,h}}(t) - DA(\mathbf{v}(t))\mathbf{w}(t)\mathbf{z_{v_k,h}}(t) \\
&\quad + D^2A(\mathbf{v}_k(t))(\mathbf{h}(t),\mathbf{w}(t))\mathbf{y_{v_k}}(t) - D^2A(\mathbf{v}(t))(\mathbf{h}(t),\mathbf{w}(t))\mathbf{y_v}(t) \\
&\quad + D^2A(\mathbf{v}(t))(\mathbf{h}(t),\mathbf{w}(t))\mathbf{y_{v_k}}(t) \\
&\quad - D^2A(\mathbf{v}(t))(\mathbf{h}(t),\mathbf{w}(t))\mathbf{y_{v_k}}(t).
\end{aligned} \tag{6.82}$$

Using similar arguments as before and Gronwall's Lemma, we obtain the estimate

$$\|\mathbf{q}_{\mathbf{v}_k,\mathbf{h},\mathbf{w}}(t) - \mathbf{q}_{\mathbf{v},\mathbf{h},\mathbf{w}}(t)\| \leq \gamma(t) \exp\Big(\int_0^T \|A(\mathbf{v}_k(s))\|_2 \, \mathrm{d}s\Big),$$

where

$$\begin{aligned}
\gamma(t) = \int_0^t \Big(& \|A(\mathbf{v}_k(s)) - A(\mathbf{v}(s))\|_2 \|\mathbf{q}_{\mathbf{v},\mathbf{h},\mathbf{w}}(s)\| \\
& + \|DA(\mathbf{v}_k(s))\mathbf{h}(s) - DA(\mathbf{v}(s))\mathbf{h}(s)\|_2 \|\mathbf{z}_{\mathbf{v}_k,\mathbf{w}}(s)\| \\
& + \|DA(\mathbf{v}(s))\mathbf{h}(s)\|_2 \|\mathbf{z}_{\mathbf{v}_k,\mathbf{w}}(s) - \mathbf{z}_{\mathbf{v},\mathbf{w}}(s)\| \\
& + \|DA(\mathbf{v}_k(s))\mathbf{w}(s) - DA(\mathbf{v}(s))\mathbf{w}(s)\|_2 \|\mathbf{z}_{\mathbf{v}_k,\mathbf{h}}(s)\| \\
& + \|DA(\mathbf{v}(s))\mathbf{w}(s)\|_2 \|\mathbf{z}_{\mathbf{v}_k,\mathbf{h}}(s) - \mathbf{z}_{\mathbf{v},\mathbf{h}}(s)\| \\
& + \|D^2 A(\mathbf{v}_k(s))(\mathbf{w}(s),\mathbf{h}(s)) - D^2 A(\mathbf{v}(s))(\mathbf{w}(s),\mathbf{h}(s))\|_2 \|\mathbf{y}_{\mathbf{v}_k}(s)\| \\
& + \|D^2 A(\mathbf{v}(s))(\mathbf{w}(s),\mathbf{h}(s))\|_2 \|\mathbf{y}_{\mathbf{v}}(s) - \mathbf{y}_{\mathbf{v}_k}(s)\| \Big) \, \mathrm{d}s.
\end{aligned}$$

We can derive the estimate

$$\begin{aligned}
\gamma(t) \leq \gamma(T) \leq c\|\mathbf{h}\|_{C^0} \|\mathbf{w}\|_{C^0} T \Big(& \|A \circ \mathbf{v}_k - A \circ \mathbf{v}\|_{C^0} + \|DA(\mathbf{v}_k) - DA(\mathbf{v})\|_{C^0} \\
& + \|DA(\mathbf{v})\|_{C^0} \|D\mathcal{S}_2(\mathbf{v}_k) - D\mathcal{S}_2(\mathbf{v})\|_{C^0} \\
& + \|D^2 A(\mathbf{v}_k) - D^2 A(\mathbf{v})\|_{C^0} \|\mathbf{y}_{\mathbf{v}_k}\|_{C^0} \\
& + \|D^2 A(\mathbf{v})\|_{C^0} \|\mathbf{y}_{\mathbf{v}} - \mathbf{y}_{\mathbf{v}_k}\|_{C^0} \Big),
\end{aligned}$$

for some constant $c > 0$, due to the continuity of $\mathbf{z}_{\mathbf{v}_k,\mathbf{w}}, \mathbf{z}_{\mathbf{v}_k,\mathbf{h}}$ with respect to $\mathbf{w}, \mathbf{h}$ and because we have $\|\mathbf{q}_{\mathbf{v},\mathbf{h},\mathbf{w}}\|_{C^1} \leq \tilde{c}\|\mathbf{h}\|_{C^0}\|\mathbf{w}\|_{C^0}$.

We can use

$$\|A \circ \mathbf{v}_k - A \circ \mathbf{v}\|_{C^0} \to 0, \tag{6.83}$$

$$\|DA(\mathbf{v}_k) - DA(\mathbf{v})\|_{C^0} \to 0, \tag{6.84}$$

$$\|D^2 A(\mathbf{v}_k) - D^2 A(\mathbf{v})\|_{C^0} \to 0, \tag{6.85}$$

$$\|D\mathcal{S}_2(\mathbf{v}_k) - D\mathcal{S}_2(\mathbf{v})\|_{C^0} \to 0, \tag{6.86}$$

$$\|\mathbf{y}_{\mathbf{v}_k} - \mathbf{y}_{\mathbf{v}}\|_{C^0} \to 0, \tag{6.87}$$

to obtain $\|D^2\mathcal{S}_2(\mathbf{v}_k) - D^2\mathcal{S}_2(\mathbf{v})\|_{C^0} \to 0$. Using this, (6.82) and (6.83)–(6.87), we can show $\|D^2\mathcal{S}_2(\mathbf{v}_k) - D^2\mathcal{S}_2(\mathbf{v})\|_{C^1} \to 0$ by similar arguments. This finishes the proof. $\square$

7 Lagrange framework and first-order necessary optimality conditions

In this chapter we will derive first-order necessary optimality conditions for the optimization problem (5.19). First, we will prove the existence of Lagrange multipliers fulfilling the first-order necessary optimality conditions by using an auxiliary system. Then we will show the existence of Lagrange multipliers for the original problem. For this purpose, we present a transformation for the Stokes–Brinkman equation (5.20)–(5.22).

Remark 7.1. From this chapter on, we have to drop the SUPG terms in (5.18). This decision is explained later in the chapter. Note that the discretized advection-diffusion equation without any upwinding terms can then be written in the form

$$M_{ad}\frac{\mathrm{d}}{\mathrm{d}t}\mathbf{c}(t) = A_{ad}(\mathbf{v}(t))\mathbf{c}(t),$$
$$M_{ad}\mathbf{c}(0) = \mathbf{c}_0,$$

where A_{ad} is a linear map. Note that we deviate in notation from Chapter 5, where the matrices were called "M_1" and "A_1", respectively. Considering all remaining terms in (5.18), it is easy to see, that $A_{ad}(\mathbf{v}(t)) = A_0 + \sum_{i=1}^{n_v} \alpha_i(t)A_i$ for some matrices $A_i \in \mathbb{R}^{n_c \times n_c}$ and $i = 0, \ldots, n_v$.

7.1 Preparations

Before we start with the main content, we will prove some lemmas which will be very useful later in this chapter. We discuss the existence of weak solutions of ordinary differential equations (ODEs) extending standard arguments in the theory of ODEs.

Lemma 7.2. *Let $\mathbf{f} \in L^2(0,T;\mathbb{R}^n)$ be given. Then the function $\mathbf{F} \in L^2(0,T;\mathbb{R}^n)$, where $\mathbf{F}(t) = \int_0^t \mathbf{f}(s)\,\mathrm{d}s$, admits a weak derivative and $\frac{\mathrm{d}}{\mathrm{d}t}\mathbf{F} = \mathbf{f}$ holds true in the weak sense.*

Proof. Clearly, $\mathbf{F}$ is a L^2-function, due to

$$\int_0^T \|\mathbf{F}(t)\|^2\,\mathrm{d}t = \int_0^T \left\| \int_0^t \mathbf{f}(s)\,\mathrm{d}s \right\|^2 \mathrm{d}t \le \int_0^T T\|\mathbf{f}\|_{L^2}^2\,\mathrm{d}t = T\|\mathbf{f}\|_{L^2}^2.$$

Next, we define $\mathcal{X}_t : [0,T] \to \{0,1\}$ as

$$\mathcal{X}_t(s) = \begin{cases} 1, & s \le t \\ 0, & s > t. \end{cases}$$

Let $\varphi \in C_0^\infty(0, T; \mathbb{R}^n)$ be arbitrarily given. Then, using the Fubini–Tonelli theorem [29] we have

$$
\begin{aligned}
\int_0^T \mathbf{F}(t) \frac{\mathrm{d}}{\mathrm{d}t} \varphi(t) \, \mathrm{d}t &= \int_0^T \int_0^t \mathbf{f}(s) \, \mathrm{d}s \, \frac{\mathrm{d}}{\mathrm{d}t} \varphi(t) \, \mathrm{d}t \\
&= \int_0^T \int_0^T \mathbf{f}(s) \, \mathcal{X}_t(s) \, \mathrm{d}s \, \frac{\mathrm{d}}{\mathrm{d}t} \varphi(t) \, \mathrm{d}t \\
&= \int_0^T \int_0^T \mathbf{f}(s) \, \mathcal{X}_t(s) \, \frac{\mathrm{d}}{\mathrm{d}t} \varphi(t) \, \mathrm{d}t \, \mathrm{d}s \\
&= \int_0^T \int_s^T \mathbf{f}(s) \, \frac{\mathrm{d}}{\mathrm{d}t} \varphi(t) \, \mathrm{d}t \, \mathrm{d}s \\
&= \int_0^T \mathbf{f}(s) \int_s^T \frac{\mathrm{d}}{\mathrm{d}t} \varphi(t) \, \mathrm{d}t \, \mathrm{d}s \\
&= \int_0^T \mathbf{f}(s) (\underbrace{\varphi(T)}_{=0} - \varphi(s)) \, \mathrm{d}s \\
&= -\int_0^T \mathbf{f}(s) \varphi(s) \, \mathrm{d}s,
\end{aligned}
$$

which shows the claim. $\qquad\square$

Lemma 7.3. *The equation*

$$
\frac{\mathrm{d}}{\mathrm{d}t} \mathbf{y}(t) = A\mathbf{y}(t) + \mathbf{f}(t), \quad \mathbf{y}(0) = \mathbf{y}_0, \tag{7.1}
$$

where $A \in \mathbb{R}^{n \times n}, \mathbf{y}_0 \in \mathbb{R}^n$ and $\mathbf{f} \in L^2(0, T; \mathbb{R}^n)$ are given, admits a unique solution $\mathbf{y} \in H^1(0, T; \mathbb{R}^n)$.

Proof. A solution of (7.1) can be obtained explicitly as

$$
\mathbf{y}(t) = e^{At} \mathbf{y}_0 + e^{At} \int_0^t e^{-As} \mathbf{f}(s) \, \mathrm{d}s. \tag{7.2}
$$

This expression is well-defined due to $\mathbf{f}$ being an $L^2(0, T; \mathbb{R}^n)$-function. Next, we prove that $\mathbf{y}$ has a weak derivative which fulfills the differential equation. Obviously, the first summand of $\mathbf{y}$ in (7.2) is arbitrarily often differentiable and the weak derivative equals the strong derivative with respect to t. For the second summand, note that $e^{-At} \mathbf{f}(t)$ is the weak derivative of $\int_0^t e^{-As} \mathbf{f}(s) \, \mathrm{d}s$ due to Lemma 7.2.

Furthermore, the function $\mathbf{y}$ belongs to $H^1(0,T;\mathbb{R}^n)$. The first summand $e^{At}\mathbf{y}_0$ is clearly an $H^1(0,T;\mathbb{R}^n)$-function. For the second summand, we obtain using the Hölder inequality that

$$\int_0^T \underbrace{\left|\int_0^t e^{-As}\mathbf{f}(s)\,\mathrm{d}s\right|^2}_{\leq \|e^{-A(\cdot)}\|^2_{L^2(0,t;\mathbb{R}^n)}\|\mathbf{f}\|^2_{L^2(0,t;\mathbb{R}^n)}} \mathrm{d}t \leq \int_0^T \|e^{-A(\cdot)}\|^2_{L^2(0,t;\mathbb{R}^n)}\|\mathbf{f}\|^2_{L^2(0,t;\mathbb{R}^n)}\,\mathrm{d}t$$

$$\leq \int_0^T \|e^{-A(\cdot)}\|^2_{L^2(0,T;\mathbb{R}^n)}\|\mathbf{f}\|^2_{L^2(0,T;\mathbb{R}^n)}\,\mathrm{d}t$$

$$= T\|e^{-A(\cdot)}\|^2_{L^2(0,T;\mathbb{R}^n)}\|\mathbf{f}\|^2_{L^2(0,T;\mathbb{R}^n)} < \infty,$$

which shows that $\mathbf{y} \in L^2(0,T;\mathbb{R}^n)$.

Finally, the weak derivative of $\mathbf{y}$ is

$$\frac{\mathrm{d}}{\mathrm{d}t}\mathbf{y}(t) = Ae^{At}\mathbf{y}_0 + Ae^{At}\int_0^t e^{-As}\mathbf{f}(s)\,\mathrm{d}s + e^{At}e^{-At}\mathbf{f}(t) = A\mathbf{y}(t) + \mathbf{f}(t),$$

which is clearly an $L^2(0,T;\mathbb{R}^n)$-function.

For the uniqueness, we assume that ϕ is a solution of the differential equation (7.1). Set $\mathbf{g}(t) := e^{-At}\phi(t)$. Then it holds

$$\frac{\mathrm{d}}{\mathrm{d}t}\mathbf{g}(t) = -Ae^{-At}\phi(t) + e^{-At}\frac{\mathrm{d}}{\mathrm{d}t}\phi(t)$$

$$= -Ae^{-At}\phi(t) + e^{-At}(A\phi(t) + \mathbf{f}(t))$$

$$= -Ae^{-At}\phi(t) + Ae^{-At}\phi(t) + e^{-At}\mathbf{f}(t)$$

$$= e^{-At}\mathbf{f}(t).$$

Thus, we have

$$\mathbf{g}(t) = \int_0^t e^{-As}\mathbf{f}(s)\,\mathrm{d}s + \mathbf{c},$$

and multiplying with e^{At} we obtain

$$\phi(t) = e^{At}\mathbf{c} + e^{At}\int_0^t e^{-As}\mathbf{f}(s)\,\mathrm{d}s$$

for the solution ϕ. Due to the initial condition, the solution is unique. $\qquad\square$

Lemma 7.4. *Let $A \in L^2(0,T;\mathbb{R}^{n\times n})$ and $\mathbf{y}_0 \in \mathbb{R}^n$ be given. Then the initial value problem*

$$\frac{\mathrm{d}}{\mathrm{d}t}\mathbf{y}(t) = A(t)\mathbf{y}(t), \tag{7.3}$$

$$\mathbf{y}(0) = \mathbf{y}_0 \tag{7.4}$$

admits a unique solution $\mathbf{y} \in H^1(0,T;\mathbb{R}^n)$.

Proof. Firstly, we consider an extension of A,

$$\tilde{A}(t) := \begin{cases} A(t), & t \leq T \\ 0, & t > T \end{cases}$$

which is a measurable function and furthermore an $L^2(0, \infty)$-function. In addition, we define a constant c as $c := \|A\|_{L^2(0,T)} = \|\tilde{A}\|_{L^2(0,\infty)}$.

Using this we prove that there exists a unique solution of (7.3)–(7.4). Consider the integral equation

$$\mathbf{y}(t) = \mathbf{y}_0 + \int_0^t A(s)\mathbf{y}(s)\,\mathrm{d}s,$$

which is equivalent to (7.3)–(7.4). We define the operator $\mathcal{T} : L^2(0, \bar{T}; \mathbb{R}^n) \to L^2(0, \bar{T}; \mathbb{R}^n)$ as

$$\mathcal{T}(\mathbf{y})(t) = \mathbf{y}_0 + \int_0^t A(s)\mathbf{y}(s)\,\mathrm{d}s,$$

where $\bar{T} \leq T$ is chosen in such a way that $\sqrt{\bar{T}}c < 1$ is fulfilled. Note that this operator is well-defined, because A and $\mathbf{y}$ are L^2-functions.

The idea is to imitate the proof of the classical Picard–Lindelöf theorem [18] using the Banach fixed point theorem [5].

We start with the estimate

$$\begin{aligned}
\|\mathcal{T}(\mathbf{y}) - \mathcal{T}(\mathbf{x})\|_{L^2} &= \left\| \int_0^t A(s)(\mathbf{y}(s) - \mathbf{x}(s))\,\mathrm{d}s \right\|_{L^2} \\
&= \left(\int_0^{\bar{T}} \left\| \int_0^t A(s)(\mathbf{y}(s) - \mathbf{x}(s))\,\mathrm{d}s \right\|^2 \mathrm{d}t \right)^{\frac{1}{2}} \\
&\leq \left(\int_0^{\bar{T}} \left(\int_0^{\bar{T}} \|A(s)\| \|\mathbf{y}(s) - \mathbf{x}(s)\|\,\mathrm{d}s \right)^2 \mathrm{d}t \right)^{\frac{1}{2}} \\
&\leq \left(\int_0^{\bar{T}} \|A\|_{L^2}^2 \|\mathbf{y} - \mathbf{x}\|_{L^2}^2\,\mathrm{d}t \right)^{\frac{1}{2}} \\
&\leq \sqrt{\bar{T}}\|A\|_{L^2} \|\mathbf{y} - \mathbf{x}\|_{L^2} \\
&= \sqrt{\bar{T}}c \|\mathbf{y} - \mathbf{x}\|_{L^2}.
\end{aligned}$$

Since we haven chosen $\bar{T}$ sufficiently small, this estimate shows that $\mathcal{T}$ is Lipschitz continuous with a Lipschitz constant less than 1. Applying the Banach fixed point theorem, we see that there exists a unique fixpoint $\varphi \in L^2(0, \bar{T}; \mathbb{R}^n)$ with

$$\varphi(t) = \mathbf{y}_0 + \int_0^t A(s)\varphi(s)\,\mathrm{d}s. \tag{7.5}$$

Due to Lemma 7.2 the right hand side of (7.5) is weakly differentiable with weak derivative given by $A(t)\varphi(t)$, which is clearly an L^2-function. Therefore φ is in $H^1(0, \bar{T}; \mathbb{R}^n)$ and a solution of the initial value problem (7.3)–(7.4).

This shows the existence of a local, unique solution on the interval $[0, \bar{T}]$. It remains to prove that there exists a unique solution on the whole time interval $[0, T]$. We define $\Delta := \bar{T}$ as the length of the interval on which we have already shown the existence.

We consider the auxiliary initial value problem on the interval $[\bar{T}, T]$

$$\frac{\mathrm{d}}{\mathrm{d}t}\mathbf{y}(t) = A(t)\mathbf{y}(t),$$
$$\mathbf{y}(\bar{T}) = \varphi(\bar{T}).$$

This problem is well-defined, due to φ being a C^0-function.

We can use the argument from above using the Banach fixed point theorem to show that this problem also admits a unique solution ψ on some interval $[\bar{T}, \hat{T}]$, where $\hat{T} = \bar{T} + \Delta$, due to

$$\|\bar{\mathcal{T}}(\mathbf{y}) - \bar{\mathcal{T}}(\mathbf{x})\|_{L^2} \le \sqrt{\hat{T} - \bar{T}}\, c \|\mathbf{y} - \mathbf{x}\|_{L^2},$$

where

$$\bar{\mathcal{T}}(y)(t) := \varphi(t) + \int_{\bar{T}}^{t} A(s)\mathbf{y}(s)\,\mathrm{d}s.$$

Now we can construct a solution on $[0, \hat{T}]$ by

$$\phi := \begin{cases} \varphi, & t \in [0, \bar{T}] \\ \psi, & t \in (\bar{T}, \hat{T}]. \end{cases}$$

This function is clearly in C^0 and is also weakly differentiable, which can be shown using the definition

$$\phi' := \begin{cases} \frac{\mathrm{d}}{\mathrm{d}t}\varphi, & t \in [0, \bar{T}] \\ \frac{\mathrm{d}}{\mathrm{d}t}\psi, & t \in (\bar{T}, \hat{T}]. \end{cases}$$

Let $\mathbf{f} \in C_0^\infty(0, \hat{T}; \mathbb{R}^n)$ be arbitrarily given. Then it holds

$$\int_0^{\hat{T}} \phi \cdot \frac{\mathrm{d}}{\mathrm{d}t}\mathbf{f}\,\mathrm{d}t = \int_0^{\bar{T}} \varphi \cdot \frac{\mathrm{d}}{\mathrm{d}t}\mathbf{f}\,\mathrm{d}t + \int_{\bar{T}}^{\hat{T}} \psi \cdot \frac{\mathrm{d}}{\mathrm{d}t}\mathbf{f}\,\mathrm{d}t$$

$$= -\int_0^{\bar{T}} \frac{\mathrm{d}}{\mathrm{d}t}\varphi \cdot \mathbf{f}\,\mathrm{d}t + (\varphi(\bar{T}), \mathbf{f}(\bar{T}))_2 - (\varphi(0), \mathbf{f}(0))_2$$

$$\quad - \int_{\bar{T}}^{\hat{T}} \frac{\mathrm{d}}{\mathrm{d}t}\psi \cdot \mathbf{f}\,\mathrm{d}t + (\psi(\hat{T}), \mathbf{f}(\hat{T}))_2 - (\psi(\bar{T}), \mathbf{f}(\bar{T}))_2$$

$$= -\int_0^{\hat{T}} \frac{\mathrm{d}}{\mathrm{d}t}\phi \cdot \mathbf{f}\,\mathrm{d}t$$

using integration by parts. The derivative is also clearly an L^2-function and furthermore the original initial value problem (7.3)–(7.4) is fulfilled.

Continuing in this way, we can construct a unique solution on the whole interval $[0, T]$. $\square$

7.2 Introduction of an auxiliary system

Before we consider the optimal control problem from the previous chapter, we turn back on the idea of the system transformation from the proof of Theorem 6.4. We start with the Stokes–Brinkman equations

$$M\frac{\mathrm{d}}{\mathrm{d}t}\mathbf{v}(t) = A\mathbf{v}(t) + B^T\mathbf{p}(t) + F_0\mathbf{u}(t) + F_1\frac{\mathrm{d}}{\mathrm{d}t}\mathbf{u}(t), \tag{7.6}$$

$$0 = B\mathbf{v}(t) + L\mathbf{u}(t), \tag{7.7}$$

$$\mathbf{z}(t) = C_1\mathbf{v}(t) \tag{7.8}$$

$$M\mathbf{v}(0) = \mathbf{v}_0, \tag{7.9}$$

where $\mathbf{v}(t) \in \mathbb{R}^{n_v}$ and $\mathbf{p}(t) \in \mathbb{R}^{n_p}$ are the state variables, $\mathbf{u}(t) \in \mathbb{R}^2$ is the input and $\mathbf{z}(t) \in \mathbb{R}^{n_v}$ is the output, respectively. Furthermore $M, A \in \mathbb{R}^{n_v \times n_v}$, $B \in \mathbb{R}^{n_p \times n_v}$, $C_1 \in \mathbb{R}^{n_v \times n_v}$, $F_0, F_1 \in \mathbb{R}^{n_v \times 2}$, $L \in \mathbb{R}^{n_p \times 2}$ and the initial condition $\mathbf{v}_{0,h} \in \mathbb{R}^{n_v}$.

Remark 7.5. For the application to field flow fractionation, the velocity $\mathbf{v}$ is required as input for the advection-diffusion equation (5.23)–(5.24). This leads to the choice $C_1 = I_{n_v} \in \mathbb{R}^{n_v \times n_v}$ with I_{n_v} being the identity matrix. However we consider the general case for an arbitrary matrix $C_1 \in \mathbb{R}^{n_v \times n_v}$ in this section.

Equations (7.6) and (7.7) can also be written using block matrices

$$\begin{pmatrix} M & 0 \\ 0 & 0 \end{pmatrix}\frac{\mathrm{d}}{\mathrm{d}t}\begin{pmatrix} \mathbf{v} \\ \mathbf{p} \end{pmatrix} = \begin{pmatrix} A & B^T \\ B & 0 \end{pmatrix}\begin{pmatrix} \mathbf{v} \\ \mathbf{p} \end{pmatrix} + \begin{pmatrix} F_0 & F_1 \\ L & 0 \end{pmatrix}\begin{pmatrix} \mathbf{u} \\ \frac{\mathrm{d}}{\mathrm{d}t}\mathbf{u} \end{pmatrix}.$$

Again, like in the proof of Theorem 6.4, we decompose the solution $\mathbf{v}$ using the projection matrix

$$\Pi = I - B^T(BM^{-1}B^T)^{-1}BM^{-1} \in \mathbb{R}^{n_v \times n_v}.$$

The solution $\mathbf{v}(t)$ of the equations (7.6)–(7.9) is decomposed as

$$\mathbf{v}(t) = \Pi^T\mathbf{v}(t) + (\mathbf{v}(t) - \Pi^T\mathbf{v}(t)) =: \mathbf{v}_1(t) + \mathbf{v}_2(t),$$

where

$$\begin{aligned} \mathbf{v}_2(t) &= \mathbf{v}(t) - \Pi^T\mathbf{v}(t) = M^{-1}B^T(BM^{-1}B^T)^{-1}B\mathbf{v}(t) \\ &= -M^{-1}B^T(BM^{-1}B^T)^{-1}L\mathbf{u}(t). \end{aligned}$$

Putting the decomposition $\mathbf{v}(t) := \mathbf{v}_1(t) + \mathbf{v}_2(t)$ into (7.6)–(7.9) we obtain

$$M\frac{\mathrm{d}}{\mathrm{d}t}\mathbf{v}_1(t) = A\mathbf{v}_1(t) + B^T\mathbf{p}(t) + (F_0 - AM^{-1}B^T(BM^{-1}B^T)^{-1}L)\mathbf{u}(t)$$
$$+ (F_1 + B^T(BM^{-1}B^T)^{-1}L)\frac{\mathrm{d}}{\mathrm{d}t}\mathbf{u}(t), \tag{7.10}$$

$$B\mathbf{v}_1(t) = \mathbf{0}, \tag{7.11}$$

$$\mathbf{z}(t) = C_1\mathbf{v}_1(t) - C_1M^{-1}B^T(BM^{-1}B^T)^{-1}L\mathbf{u}(t), \tag{7.12}$$

$$M\mathbf{v}_1(0) = \Pi\mathbf{v}_0. \tag{7.13}$$

Similar to the proof of Theorem 6.4 we obtain the explicit expression

$$\mathbf{p}(t) = -(BM^{-1}B^T)^{-1}BM^{-1}A\mathbf{v}_1(t)$$
$$- (BM^{-1}B^T)^{-1}BM^{-1}(F_0 - AM^{-1}B^T(BM^{-1}B^T)^{-1}L)\mathbf{u}(t) \tag{7.14}$$
$$- (BM^{-1}B^T)^{-1}(L + BM^{-1}F_1)\frac{\mathrm{d}}{\mathrm{d}t}\mathbf{u}(t)$$

for the pressure. Again, we multiply the equations (7.10)–(7.13) from the left with Π, use $\Pi^T\mathbf{v}_1(t) = \mathbf{v}_1(t)$ and (7.14) in order to obtain

$$\Pi M\Pi^T\frac{\mathrm{d}}{\mathrm{d}t}\mathbf{v}_1(t) = \Pi A\Pi^T\mathbf{v}_1(t) + \Pi\bar{B}\mathbf{u}(t) + \Pi F_1\frac{\mathrm{d}}{\mathrm{d}t}\mathbf{u}(t), \tag{7.15}$$

$$\Pi^T\mathbf{v}_1(t) = \mathbf{v}_1(t), \tag{7.16}$$

$$\mathbf{z}(t) = \bar{C}\Pi^T\mathbf{v}_1(t) + \bar{D}\mathbf{u}(t), \tag{7.17}$$

$$\Pi M\Pi^T\mathbf{v}_1(0) = \Pi\mathbf{v}_0, \tag{7.18}$$

where the matrices $\bar{B}, \bar{C}$ and $\bar{D}$ are given as

$$\bar{B} := F_0 - AM^{-1}B^T(BM^{-1}B^T)^{-1}L,$$
$$\bar{C} := C_1,$$
$$\bar{D} := -C_1M^{-1}B^T(BM^{-1}B^T)^{-1}L.$$

In the last step, we introduce again the factorization of the projector matrix Π, using two matrices $\Theta_l, \Theta_r \in \mathbb{R}^{n_v \times (n_v - n_p)}$, which is given by

$$\Pi = \Theta_l\Theta_r^T, \text{ where}$$
$$\Theta_l^T\Theta_r = I.$$

Putting this representation of Π into (7.15)–(7.18), multiplying from the left with Θ_r^T and using $\bar{\mathbf{v}}_1(t) := \Theta_l^T\mathbf{v}_1(t) \in \mathbb{R}^{n_v - n_p}$ we finally arrive at

$$\Theta_r^T M\Theta_r\frac{\mathrm{d}}{\mathrm{d}t}\bar{\mathbf{v}}_1(t) = \Theta_r^T A\Theta_r\bar{\mathbf{v}}_1(t) + \Theta_r^T\bar{B}\mathbf{u}(t) + \Theta_r^T F_1\frac{\mathrm{d}}{\mathrm{d}t}\mathbf{u}(t), \tag{7.19}$$

$$\mathbf{z}(t) = \bar{C}\Theta_r\bar{\mathbf{v}}_1(t) + \bar{D}\mathbf{u}(t), \tag{7.20}$$

$$\Theta_r^T M\Theta_r\bar{\mathbf{v}}_1(0) = \Theta_r^T\mathbf{v}_0. \tag{7.21}$$

We emphasize the fact, that for a given input $\mathbf{u}$, the systems (7.6)–(7.9) and (7.19)–(7.21) have the same output $\mathbf{z}$.

7.3 Existence of Lagrange multipliers

We want to establish the first-order necessary optimality conditions using the method of Lagrange multipliers [41]. In order to show the existence of Lagrange multipliers, we consider a modified optimal control problem. Thereby, the transformations from the previous section will be important. We start with the following optimal control problem

$$\min J(\mathbf{u}) = \frac{1}{2}\|\mathbf{c}(T) - \mathbf{c}^{\text{foc}}\|_{M_{ad}}^2 + \frac{\sigma}{2}\|\mathbf{u}\|_{L^2(0,T;\mathbb{R}^2)}^2 \tag{7.22}$$

governed by the conditions

$$\Theta_r^T M \Theta_r \frac{\mathrm{d}}{\mathrm{d}t}\bar{\mathbf{v}}_1(t) - \Theta_r^T A \Theta_r \bar{\mathbf{v}}_1(t) - \Theta_r^T \bar{B}\mathbf{u}(t) - \Theta_r^T F_1 \frac{\mathrm{d}}{\mathrm{d}t}\mathbf{u}(t) = 0, \tag{7.23}$$

$$\Theta_r^T M \Theta_r \bar{\mathbf{v}}_1(0) - \Theta_r^T \mathbf{v}_0 = 0, \tag{7.24}$$

$$\mathbf{z}(t) - \bar{C}\Theta_r \bar{\mathbf{v}}_1(t) - \bar{D}\mathbf{u}(t) = 0, \tag{7.25}$$

$$M_{ad}\frac{\mathrm{d}}{\mathrm{d}t}\mathbf{c}(t) - A_{ad}(\mathbf{z}(t))\mathbf{c}(t) = 0, \tag{7.26}$$

$$M_{ad}\mathbf{c}(0) = \mathbf{c}_0, \tag{7.27}$$

where $\bar{B} = F_0 - AM^{-1}B^T(BM^{-1}B^T)^{-1}L$, $\bar{C} = C_1$, $\bar{D} = -C_1 M^{-1}B^T(BM^{-1}B^T)^{-1}L$. Now we consider the mapping $\mathcal{H} : \Upsilon \times \mathcal{X} \to \mathcal{Y}$ with

$$\mathcal{X} = H^1(0,T;\mathbb{R}^{n_v-n_p}) \times L^2(0,T;\mathbb{R}^{n_v}) \times H^1(0,T;\mathbb{R}^{n_c}) \tag{7.28}$$

$$\mathcal{Y} = L^2(0,T;\mathbb{R}^{n_v-n_p}) \times \mathbb{R}^{n_v-n_p} \times L^2(0,T;\mathbb{R}^{n_v}) \times L^2(0,T;\mathbb{R}^{n_c}) \times \mathbb{R}^{n_c} \tag{7.29}$$

which is defined in the following way:

$$\mathcal{H}(\mathbf{u},\bar{\mathbf{v}},\mathbf{z},\mathbf{c}) = \begin{pmatrix} \Theta_r^T M \Theta_r \frac{\mathrm{d}}{\mathrm{d}t}\bar{\mathbf{v}} - \Theta_r^T A \Theta_r \bar{\mathbf{v}} - \Theta_r^T \bar{B}\mathbf{u} - \Theta_r^T F_1 \frac{\mathrm{d}}{\mathrm{d}t}\mathbf{u} \\ \Theta_r^T M \Theta_r \bar{\mathbf{v}}(0) - \Theta_r^T \mathbf{v}_0 \\ \mathbf{z} - \bar{C}\Theta_r \bar{\mathbf{v}} - \bar{D}\mathbf{u} \\ M_{ad}\frac{\mathrm{d}}{\mathrm{d}t}\mathbf{c} - A_{ad}(\mathbf{z})\mathbf{c} \\ M_{ad}\mathbf{c}(0) - \mathbf{c}_0 \end{pmatrix}$$

One main goal of this section is to prove the following two theorems: The first one gives the first-order optimality conditions which are important for the theory. The second one gives an explicit description of the gradient of J which is used in numerical optimization methods.

We use the notation $f(\mathbf{u},\mathbf{x}) = f(\mathbf{u},(\bar{\mathbf{v}},\mathbf{z},\mathbf{c})) = \frac{1}{2}\|\mathbf{c}(T) - \mathbf{c}^{\text{foc}}\|_{M_{ad}}^2 + \frac{\sigma}{2}\|\mathbf{u}\|_{L^2(0,T;\mathbb{R}^2)}^2$, then the optimization problem (7.22) can be concisely formulated as

$$\min f(\mathbf{u},\mathbf{x}) \quad \text{where} \quad \mathcal{H}(\mathbf{u},\mathbf{x}) = 0.$$

Theorem 7.6 (First-oder necessary optimality conditions). *Let $\mathbf{u} \in \Upsilon$ be a local minimum of $J : \Upsilon \to \mathbb{R}$ and let $\mathbf{x} \in \mathcal{X}$ be the unique vector satisfying $\mathcal{H}(\mathbf{u}, \mathbf{x}) = 0$. Then there exists a vector $\Lambda \in \mathcal{Y}$ such that*

$$D_{\mathbf{x}} f(\mathbf{u}, \mathbf{x}) \mathbf{h} + (\Lambda, D_{\mathbf{x}} \mathcal{H}(\mathbf{u}, \mathbf{x}) \mathbf{h})_{\mathcal{Y}} = 0 \quad \text{for all } \mathbf{h} \in \mathcal{X}, \tag{7.30}$$

$$D_{\mathbf{u}} f(\mathbf{u}, \mathbf{x}) \mathbf{h} + (\Lambda, D_{\mathbf{u}} \mathcal{H}(\mathbf{u}, \mathbf{x}) \mathbf{h})_{\mathcal{Y}} = 0 \quad \text{for all } \mathbf{h} \in \Upsilon. \tag{7.31}$$

Here, Λ is a *Lagrange multiplier*.

Theorem 7.7 (Gradient representation). *Let $\mathbf{u} \in \Upsilon$ be an arbitrary vector. Let $\mathbf{x} \in \mathcal{X}$ be the unique vector such that $\mathcal{H}(\mathbf{u}, \mathbf{x}) = 0$. Then there exists a unique vector $\Lambda \in \mathcal{Y}$ satisfying*

$$D_{\mathbf{x}} f(\mathbf{u}, \mathbf{x}) \mathbf{h} + (\Lambda, D_{\mathbf{x}} \mathcal{H}(\mathbf{u}, \mathbf{x}) \mathbf{h})_{\mathcal{Y}} = 0, \quad \text{for all } \mathbf{h} \in \mathcal{X},$$

and the differential of J in $\mathbf{u}$ is given by

$$DJ(\mathbf{u}) \mathbf{h} = D_{\mathbf{u}} f(\mathbf{u}, \mathbf{x}) \mathbf{h} + (\Lambda, D_{\mathbf{u}} \mathcal{H}(\mathbf{u}, \mathbf{x}) \mathbf{h})_{\mathcal{Y}} \quad \text{for all } \mathbf{h} \in \Upsilon.$$

In order to show Theorem 7.6 and 7.7, we present the following proposition first.

Proposition 7.8. *Let $(\mathbf{u}, \mathbf{x}) \in \Upsilon \times \mathcal{X}$. Assume that $\mathcal{H}(\mathbf{u}, \mathbf{x}) = 0$. Then the map $D_{\mathbf{x}} \mathcal{H}(\mathbf{u}, \mathbf{x}) : \mathcal{X} \to \mathcal{Y}$ is invertible.*

For the proof of Proposition 7.8, we need to show that the derivative $\mathbf{h} \mapsto D\mathcal{H}_{\mathbf{x}}(\mathbf{x}) \mathbf{h}$ of the mapping $\mathcal{H}$ is bijective for all $\mathbf{x} \in \mathcal{X}$ with $\mathcal{H}(\mathbf{u}, \mathbf{x}) = 0$. Therefore we compute the derivative of $\mathcal{H}$ with respect to the variable $\mathbf{x} = (\bar{\mathbf{v}}, \mathbf{z}, \mathbf{y})$ in the direction $\mathbf{h} = (\mathbf{h}_1, \mathbf{h}_2, \mathbf{h}_3)$. We have

$$
\begin{aligned}
D\mathcal{H}_{\mathbf{x}}(\mathbf{u}, \mathbf{x}) \mathbf{h} &= D\mathcal{H}_{\mathbf{x}}(\mathbf{u}, \bar{\mathbf{v}}, \mathbf{z}, \mathbf{y})(\mathbf{h}_1, \mathbf{h}_2, \mathbf{h}_3) \\
&= \begin{pmatrix} \Theta_r^T M \Theta_r \frac{\mathrm{d}}{\mathrm{d}t} \mathbf{h}_1 - \Theta_r^T A \Theta_r \mathbf{h}_1 \\ \Theta_r^T M \Theta_r \mathbf{h}_1(0) \\ \mathbf{h}_2 - \bar{C} \Theta_r \mathbf{h}_1 \\ M_{ad} \frac{\mathrm{d}}{\mathrm{d}t} \mathbf{h}_3 - (D A_{ad}(\mathbf{z}) \mathbf{h}_2) \mathbf{c} - A_{ad}(\mathbf{z}) \mathbf{h}_3 \\ M_{ad} \mathbf{h}_3(0) \end{pmatrix}.
\end{aligned} \tag{7.32}
$$

Let $(\mathbf{y}_1, \mathbf{y}_2, \mathbf{y}_3, \mathbf{y}_4, \mathbf{y}_5) \in \mathcal{Y}$ be arbitrarily chosen. We need to prove the existence of a unique vector $(\mathbf{h}_1, \mathbf{h}_2, \mathbf{h}_3) \in \mathcal{X}$ such that the following equation is fulfilled:

$$
\begin{pmatrix} \Theta_r^T M \Theta_r \frac{\mathrm{d}}{\mathrm{d}t} \mathbf{h}_1 - \Theta_r^T A \Theta_r \mathbf{h}_1 \\ \Theta_r^T M \Theta_r \mathbf{h}_1(0) \\ \mathbf{h}_2 - \bar{C} \Theta_r \mathbf{h}_1 \\ M_{ad} \frac{\mathrm{d}}{\mathrm{d}t} \mathbf{h}_3 - (D A_{ad}(\mathbf{z}) \mathbf{h}_2) \mathbf{c} - A_{ad}(\mathbf{z}) \mathbf{h}_3 \\ M_{ad} \mathbf{h}_3(0) \end{pmatrix} = \begin{pmatrix} \mathbf{y}_1 \\ \mathbf{y}_2 \\ \mathbf{y}_3 \\ \mathbf{y}_4 \\ \mathbf{y}_5 \end{pmatrix}. \tag{7.33}
$$

Remark 7.9. This part, where the bijectivity is considered, is the reason for the use of the auxiliary system. If we start instead with the non-transformed Stokes–Brinkman equation (6.2)–(6.4), then the first equation of (7.33) will be problematic. The right-hand side $\mathbf{y}_1$ is an L^2-function, but Stokes-type equations require in general differentiability of the right-hand side, see Theorem 6.4. There are exceptions like, for example, the case where the input matrix L is equal to zero, but in the presented work, this is unfortunately not the case. At this point, we would not be able to show the bijectivity of $D_{\mathbf{x}}\mathcal{H}(\mathbf{x})$. On the other hand, the choice of the function space with respect to the first condition (6.2) in $L^2(0, T; \mathbb{R}^{n_v})$ is quite natural, because the derivative of the velocity occurs in this condition and we do not want to require higher regularity for the solution $\mathbf{v}$ of the Stokes–Brinkman equation than to be a $H^1(0, T; \mathbb{R}^{n_v})$-function.

We are finally able to prove Proposition 7.8 using the results from Section 7.1.

Proof of Proposition 7.8. At first we consider the first two equations in (7.33). They give an initial value problem for $\mathbf{h}_1$ with an initial vector $\mathbf{y}_2$ and an inhomogenity $\mathbf{f}_1 = \mathbf{y}_1$. Due to the regularity of $\Theta_r^T M \Theta_r$, it follows from Lemma 7.3 that this problem has a unique solution $\mathbf{h}_1 \in H^1(0, T; \mathbb{R}^{n_v - n_p})$ for any $\mathbf{y}_2 \in \mathbb{R}^{n_v - n_p}$ and $\mathbf{f}_1 \in L^2(0, T; \mathbb{R}^{n_v - n_p})$. From the third equation in (7.33) we determine $\mathbf{h}_2 = \mathbf{y}_3 + \bar{C}\Theta_r\mathbf{h}_1$. Finally, we consider the last two equations in (7.33). These equations have the form (7.3),(7.4) with the initial vector $\mathbf{y}_5$. and the right-hand side $\mathbf{y}_4 + (DA_{ad}(\mathbf{z})\mathbf{h}_2)\mathbf{c}$. We assumed $\mathbf{z}$ to be a L^2-function and A_{ad} and its derivative $DA_{ad}(\cdot)$ are linear. Thus, $A(\mathbf{z}), DA(\mathbf{z})$ are L^2-functions as well. Then, it is easy to see that system (7.33) admits a solution such that the existence of a unique solution $\mathbf{h}_3$ is ensured due to Lemma 7.4 and that finishes the proof. $\qquad\square$

Remark 7.10. The reason for striking the SUPG-terms in the discretization of the advection diffusion equation (7.26)–(7.27) lies also in the proof of Proposition 7.8. With $\mathbf{z}$ being a L^2-function, it would not be guaranteed, that either $A(\mathbf{z})$ or $DA(\mathbf{z})$ are L^2-functions as well due to the quadratic terms in $\mathbf{z}$ that appear in the description of $A(\mathbf{z})$.

Now we can show the two final results of this section.

Proof of Theorem 7.6. By Proposition 7.8, the map $D\mathcal{H}(\mathbf{u}, \mathbf{x}) : \Upsilon \times \mathcal{X} \to \mathcal{Y}$ is surjective. Therefore, the Lagrange multiplier theorem for Banach spaces [74] is applicable and this finishes the proof. $\qquad\square$

Proof of Theorem 7.7. The implicit function theorem on Banach spaces [71] yields a continuously differentiable map $g : \Upsilon \to \mathcal{X}$ such that

$$\mathcal{H}(\mathbf{u}, \mathbf{x}) = 0 \iff \mathbf{x} = g(\mathbf{u}) \quad \text{for all } \mathbf{u} \in \Upsilon \text{ and } \mathbf{x} \in \mathcal{X}.$$

Note that the assumptions of the implicit function theorem are fulfilled due to Proposition 7.8. Therefore, the functional J can be written as $J(\mathbf{u}) = f(\mathbf{u}, g(\mathbf{u}))$. Then we have

$$DJ(\mathbf{u})\mathbf{h} = D_{\mathbf{u}}f(\mathbf{u}, g(\mathbf{u}))\mathbf{h} + D_{\mathbf{x}}f(\mathbf{u}, g(\mathbf{u}))Dg(\mathbf{u})\mathbf{h}.$$

Differentiating the identity $\mathcal{H}(\mathbf{u}, g(\mathbf{u})) = 0$ yields the equation

$$D_{\mathbf{u}}\mathcal{H}(\mathbf{u}, g(\mathbf{u}))\mathbf{h} + D_{\mathbf{x}}\mathcal{H}(\mathbf{u}, g(\mathbf{u}))Dg(\mathbf{u})\mathbf{h} = 0 \text{ for all } \mathbf{h} \in \Upsilon,$$

for the derivative of g. It remains to connect this description with the Lagrange-theoretical formulation of the claim. Since $D_{\mathbf{x}}\mathcal{H}(\mathbf{u}, g(\mathbf{u}))$ is invertible, by the Riesz representation theorem [63] there is a unique $\mathbf{\Lambda} \in \mathcal{Y}$ such that

$$D_{\mathbf{x}}f(\mathbf{u}, g(\mathbf{u}))\mathbf{h} + (\mathbf{\Lambda}, D_{\mathbf{x}}\mathcal{H}(\mathbf{u}, g(\mathbf{u}))\mathbf{h})_{\mathcal{Y}} = 0 \text{ for all } \mathbf{h} \in \mathcal{X},$$

namely the unique $\mathbf{\Lambda}$ such that

$$(\mathbf{\Lambda}, \mathbf{q})_{\mathcal{Y}} = -D_{\mathbf{x}}f(\mathbf{u}, g(\mathbf{u}))D_{\mathbf{x}}\mathcal{H}(\mathbf{u}, g(\mathbf{u}))^{-1}\mathbf{q} \text{ for all } \mathbf{q} \in \mathcal{Y}.$$

Due to $Dg(\mathbf{u})\mathbf{h} = -D_{\mathbf{x}}\mathcal{H}(\mathbf{u}, g(\mathbf{u}))^{-1}D_{\mathbf{u}}\mathcal{H}(\mathbf{u}, g(\mathbf{u}))\mathbf{h}$ the claim follows. $\qquad\square$

7.4 The first-order optimality conditions for the auxiliary problem

In this section we consider the Lagrangian and derive the first-order optimality conditions for the optimal control problem (7.22)–(7.27). Let

$$\hat{\mathcal{X}} = \mathcal{X} \times L^2(0, T; \mathbb{R}^{n_v - n_p}) \times \mathbb{R}^{n_v - n_p} \times L^2(0, T; \mathbb{R}^{n_v}) \times L^2(0, T; \mathbb{R}^{n_c}) \times \mathbb{R}^{n_c}$$

with $\mathcal{X}$ as in (7.28). Consider the Lagrangian $\bar{\mathcal{L}} : \Upsilon \times \hat{\mathcal{X}} \to \mathbb{R}$ defined as

$$\begin{aligned}
\bar{\mathcal{L}}(\mathbf{u}, \bar{\mathbf{v}}, \mathbf{z}, \mathbf{c}, \bar{\boldsymbol{\lambda}}, \bar{\boldsymbol{\lambda}}_0, \mathbf{w}, \mathbf{d}, \mathbf{d}_0) &= \frac{1}{2}\|\mathbf{c}(T) - \mathbf{c}^{\mathrm{foc}}\|_{M_{ad}}^2 + \frac{\sigma}{2}\|\mathbf{u}\|_{L^2}^2 \\
&\quad + \int_0^T (\Theta_r^T M \Theta_r \frac{\mathrm{d}}{\mathrm{d}t}\bar{\mathbf{v}} - \Theta_r^T A \Theta_r \bar{\mathbf{v}} - \Theta_r^T \bar{B}\mathbf{u} - \Theta_r^T F_1 \frac{\mathrm{d}}{\mathrm{d}t}\mathbf{u}, \bar{\boldsymbol{\lambda}})_2 \, \mathrm{d}t \\
&\quad + (\Theta_r^T M \Theta_r \bar{\mathbf{v}}(0) - \Theta_r^T \mathbf{v}_0, \bar{\boldsymbol{\lambda}}_0)_2 + \int_0^T (\mathbf{z} - \bar{C}\Theta_r \bar{\mathbf{v}} - \bar{D}\mathbf{u}, \mathbf{w})_2 \, \mathrm{d}t \\
&\quad + \int_0^T (M_{ad} \frac{\mathrm{d}}{\mathrm{d}t}\mathbf{c} - A_{ad}(\mathbf{z})\mathbf{c}, \mathbf{d})_2 \, \mathrm{d}t + (M_{ad}\mathbf{c}(0) - \mathbf{c}_0, \mathbf{d}_0)_2.
\end{aligned}$$

The existence of the Lagrange multipliers has been shown in Section 7.3. We start computing the first-order optimality conditions.

Setting the derivative of $\bar{\mathcal{L}}$ with respect to $\bar{\mathbf{v}}$ in direction $\mathbf{h}$ to be zero leads to the condition

$$\begin{aligned}
D_{\bar{\mathbf{v}}}\bar{\mathcal{L}}(\mathbf{u}, \bar{\mathbf{v}}, \mathbf{z}, \mathbf{c}, \bar{\boldsymbol{\lambda}}, \bar{\boldsymbol{\lambda}}_0, \mathbf{w}, \mathbf{d}, \mathbf{d}_0)\mathbf{h} &= \int_0^T (\Theta_r^T M \Theta_r \frac{\mathrm{d}}{\mathrm{d}t}\mathbf{h} - \Theta_r^T A \Theta_r \mathbf{h}, \bar{\boldsymbol{\lambda}})_2 \, \mathrm{d}t \\
&\quad + \int_0^T (-\bar{C}\Theta_r \mathbf{h}, \mathbf{w})_2 \, \mathrm{d}t + (\Theta_r^T M \Theta_r \mathbf{h}(0), \bar{\boldsymbol{\lambda}}_0)_2 = 0,
\end{aligned}$$

(7.34)

which should be fulfilled for all $\mathbf{h} \in H^1(0, T; \mathbb{R}^{n_v - n_p})$.

The Lagrange multiplier $\bar{\lambda}$ is a priori a L^2-function due to our choice of the function spaces, but it can be shown that it is of higher regularity. We recall therefore Defintion 3.1 of a weak derivative in the sense of a Gelfand triple $V \hookrightarrow H \cong H^* \hookrightarrow V^*$. In this case, we choose $V = H = \mathbb{R}^{n_v - n_p}$. Equation (7.34) should be fulfilled especially for $\mathbf{h} = \phi(t)\mathbf{s}$, where $\mathbf{s} \in \mathbb{R}^{n_v - n_p}$ and $\phi \in C_0^\infty(0, T)$ is chosen arbitrarily. It yields

$$\int_0^T \left(\frac{\mathrm{d}}{\mathrm{d}t}\phi(t) \right) \cdot (\Theta_r^T M \Theta_r \mathbf{s}, \bar{\lambda})_2 \, \mathrm{d}t + \int_0^T \phi(t) \cdot (-\Theta_r^T A \Theta_r \mathbf{s}, \bar{\lambda})_2 \, \mathrm{d}t$$

$$+ \int_0^T \phi(t) \cdot (-\bar{C}\Theta_r \mathbf{s}, \mathbf{w})_2 \, \mathrm{d}t + \underbrace{\phi(0)}_{=0} \cdot (\Theta_r^T M \Theta_r \mathbf{s}, \bar{\lambda}_0)_2 = 0,$$

such that $\Theta_r^T M^T \Theta_r \bar{\lambda}$ is weakly differentiable (which ensures that also $\bar{\lambda}$ is weakly differentiable due to the regularity of $\Theta_r^T M^T \Theta_r$) and the differential equation

$$-\Theta_r^T M \Theta_r \frac{\mathrm{d}}{\mathrm{d}t}\bar{\lambda} - \Theta_r^T A^T \Theta_r \bar{\lambda} - \Theta_r^T \bar{C}^T \mathbf{w} = 0, \tag{7.35}$$

is fulfilled due to $M = M^T$. Integrating by parts, we get

$$\int_0^T (\Theta_r^T M \Theta_r \frac{\mathrm{d}}{\mathrm{d}t}\mathbf{h}, \bar{\lambda})_2 \, \mathrm{d}t = - \int_0^T (\mathbf{h}, \Theta_r^T M \Theta_r \frac{\mathrm{d}}{\mathrm{d}t}\bar{\lambda})_2 \, \mathrm{d}t$$

$$+ (\mathbf{h}(T), \Theta_r^T M \Theta_r \bar{\lambda}(T))_2 - (\mathbf{h}(0), \Theta_r^T M \Theta_r \bar{\lambda}(0)).$$

Putting this in (7.34) and taking into account (7.35) we obtain the conditions $\bar{\lambda}(T) = 0$ and $\bar{\lambda}(0) = \bar{\lambda}_0$. Thus, the Lagrange multiplier $\bar{\lambda}$ satisfies the system

$$-\Theta_r^T M \Theta_r \frac{\mathrm{d}}{\mathrm{d}t}\bar{\lambda} - \Theta_r^T A \Theta_r \bar{\lambda} - \Theta_r^T \bar{C}^T \mathbf{w} = 0, \tag{7.36}$$

$$\bar{\lambda}(T) = 0, \tag{7.37}$$

and the Lagrange multiplier $\bar{\lambda}_0$ can be computed as $\bar{\lambda}(0) = \bar{\lambda}_0$ by solving this system backwards.

Next, we compute the derivative of $\bar{\mathcal{L}}$ with respect to $\mathbf{z}$ in direction $\mathbf{h}$ and get it to be zero. This leads to the condition

$$D_{\mathbf{z}}\bar{\mathcal{L}}(\mathbf{u}, \bar{\mathbf{v}}, \mathbf{z}, \mathbf{c}, \bar{\lambda}, \bar{\lambda}_0, \mathbf{w}, \mathbf{d}, \mathbf{d}_0)\mathbf{h} = \int_0^T (\mathbf{h}, \mathbf{w})_2 \, \mathrm{d}t - \int_0^T (DA_{ad}(\mathbf{z})\mathbf{h}\mathbf{c}, \mathbf{d})_2 \, \mathrm{d}t = 0, \tag{7.38}$$

which should hold true for all $\mathbf{h} \in L^2(0, T; \mathbb{R}^{n_v})$.

Introducing the linear operator $\mathcal{F}_{\mathbf{z},\mathbf{c}} : L^2(0, T; \mathbb{R}^{n_v}) \rightarrow L^2(0, T; \mathbb{R}^{n_c})$ as

$$\mathcal{F}_{\mathbf{z},\mathbf{c}}\mathbf{h} = DA_{ad}(\mathbf{z})\mathbf{h}\mathbf{c}, \tag{7.39}$$

we obtain the identity

$$\int_0^T (DA_{ad}(\mathbf{z})\mathbf{h}\mathbf{c}, \mathbf{d})_2 \, \mathrm{d}t = (\mathcal{F}_{\mathbf{z},\mathbf{c}}\mathbf{h}, \mathbf{d})_{L^2} = (\mathbf{h}, \mathcal{F}_{\mathbf{z},\mathbf{c}}^* \mathbf{d})_{L^2},$$

where $\mathcal{F}_{\mathbf{z},\mathbf{c}}^*$ is the adjoint operator of $\mathcal{F}_{\mathbf{z},\mathbf{c}}$. Thus, it follows from (7.38) that

$$\mathbf{w} - \mathcal{F}_{\mathbf{z},\mathbf{c}}^* \mathbf{d} = 0.$$

Remark 7.11. Considering Remark 7.1, $A_{ad}(\mathbf{z})$ is of the form $A_{ad}(\mathbf{z}) = A_0 + \sum_{i=1}^{n_v} \mathbf{z}_i A_i$. Thus, $D A_{ad}(\mathbf{z})\mathbf{h} = \sum_{i=1}^{n_v} \mathbf{h}_i A_i$. Then we have

$$(D A_{ad}(\mathbf{z})\mathbf{h}\mathbf{c}, \mathbf{d})_{L^2} = ((\sum_{i=1}^{n_v} \mathbf{h}_i A_i)\mathbf{c}, \mathbf{d})_{L^2} = \int_0^T \sum_{i=1}^{n_v} \mathbf{h}_i (A_i \mathbf{c}, \mathbf{d})_2 \, \mathrm{d}t = (\mathbf{F}, \mathbf{h})_{L^2},$$

where $(\mathbf{F})_i = (A_i \mathbf{c}, \mathbf{d})_2$ for $i = 1, \ldots, n_v$. This gives an explicit description $\mathcal{F}_{\mathbf{z},\mathbf{c}}^* \mathbf{d} = \mathbf{F}$. For conciseness, we continue to write the general form $\mathcal{F}_{\mathbf{z},\mathbf{c}}^* \mathbf{d}$ instead of using the explicit formula.

The next step is to compute the derivative of $\bar{\mathcal{L}}$ with respect to $\mathbf{c}$ in direction $\mathbf{h}$. Setting it to zero, we have

$$\begin{aligned} D_{\mathbf{c}} \bar{\mathcal{L}}(\mathbf{u}, \bar{\mathbf{v}}, \mathbf{z}, \mathbf{c}, \bar{\boldsymbol{\lambda}}, \bar{\boldsymbol{\lambda}}_0, \mathbf{w}, \mathbf{d}, \mathbf{d}_0)\mathbf{h} &= \int_0^T (M_{ad} \frac{\mathrm{d}}{\mathrm{d}t}\mathbf{h} - A_{ad}(\mathbf{z})\mathbf{h}, \mathbf{d})_2 \, \mathrm{d}t \\ &\quad + (M_{ad}\mathbf{h}(0), \mathbf{d}_0)_2 + (\mathbf{h}(T), M_{ad}(\mathbf{c}(T) - \mathbf{c}^{\mathrm{foc}}))_2 \\ &= 0. \end{aligned}$$

$$(7.40)$$

We can use the same argumentation as for equation (7.34) to show the weak differentiability of the Lagrange multiplier $\mathbf{d}$. In conclusion, the weak derivative of $\mathbf{d}$ satisfies

$$-M_{ad} \frac{\mathrm{d}}{\mathrm{d}t}\mathbf{d} - (A_{ad}(\mathbf{z}))^T \mathbf{d} = 0, \qquad (7.41)$$

due to $M_{ad}^T = M_{ad}$. Integration by parts leads then to

$$\int_0^T (M_{ad} \frac{\mathrm{d}}{\mathrm{d}t}\mathbf{h}, \mathbf{d})_2 \, \mathrm{d}t = - \int_0^T (\mathbf{h}, M_{ad} \frac{\mathrm{d}}{\mathrm{d}t}\mathbf{d})_2 \, \mathrm{d}t + (\mathbf{h}(T), M_{ad}\mathbf{d}(T))_2 - (\mathbf{h}(0), M_{ad}\mathbf{d}(0))_2.$$

Putting this into (7.40) and using (7.41), we obtain the conditions

$$\mathbf{d}(T) = -(\mathbf{c}(T) - \mathbf{c}^{\mathrm{foc}}), \qquad (7.42)$$
$$\mathbf{d}(0) = \mathbf{d}_0. \qquad (7.43)$$

Finally, we compute the derivative of $\bar{\mathcal{L}}$ with respect to $\mathbf{u}$ in direction $\mathbf{h}$ and get it to be zero. Then we have

$$\begin{aligned} & D_{\mathbf{u}} \bar{\mathcal{L}}(\mathbf{u}, \bar{\mathbf{v}}, \mathbf{z}, \mathbf{c}, \bar{\boldsymbol{\lambda}}, \bar{\boldsymbol{\lambda}}_0, \mathbf{w}, \mathbf{d}, \mathbf{d}_0)\mathbf{h} \\ &= \sigma \int_0^T (\mathbf{h}, \mathbf{u})_2 \, \mathrm{d}t - \int_0^T (\Theta_r^T \bar{B}\mathbf{h} + \Theta_r^T F_1 \frac{\mathrm{d}}{\mathrm{d}t}\mathbf{h}, \bar{\boldsymbol{\lambda}})_2 \, \mathrm{d}t - \int_0^T (\bar{D}\mathbf{h}, \mathbf{w})_2 \, \mathrm{d}t \\ &= 0. \end{aligned} \qquad (7.44)$$

Integration by parts yields

$$\int_0^T (\Theta_r^T F_1 \frac{\mathrm{d}}{\mathrm{d}t}\mathbf{h}, \bar{\boldsymbol{\lambda}})_2 \, \mathrm{d}t = - \int_0^T (\mathbf{h}, F_1^T \Theta_r \frac{\mathrm{d}}{\mathrm{d}t}\bar{\boldsymbol{\lambda}})_2 \, \mathrm{d}t \qquad (7.45)$$
$$+ (\Theta_r^T F_1 \mathbf{h}(T), \bar{\boldsymbol{\lambda}}(T))_2 - (\Theta_r^T F_1 \mathbf{h}(0), \bar{\boldsymbol{\lambda}}(0))_2. \qquad (7.46)$$

Since $\mathbf{h} \in \Upsilon$, $\mathbf{h}(0) = \mathbf{0}$ and, hence, due to $\bar{\boldsymbol{\lambda}}(T) = \mathbf{0}$ the equation

$$\mathcal{P}_{\Upsilon}\left(\sigma \mathbf{u} - \bar{B}^T \Theta_r \bar{\boldsymbol{\lambda}} - \bar{D}^T \mathbf{w} + F_1^T \Theta_r \frac{\mathrm{d}}{\mathrm{d}t}\bar{\boldsymbol{\lambda}}\right) = \mathbf{0}$$

can be derived, where $\mathcal{P}_{\Upsilon} : L^2(0, T; \mathbb{R}^2) \to \Upsilon$ is the orthogonal projection of $L^2(0, T; \mathbb{R}^2)$ onto Υ.

Remark 7.12. It is possible to drop the requirement $\mathbf{s}(0) = \mathbf{0}$ for $\mathbf{s} \in \Upsilon$. In this case one obtains an additional term in the gradient, as then one of the scalar products in (7.46) does not vanish.

Theorem 7.13. *Let $\mathbf{u} \in \Upsilon$ be an arbitrary vector. Then*

$$\nabla J(\mathbf{h}) = \mathcal{P}_{\Upsilon}\left(\sigma \mathbf{u} - \bar{B}^T \Theta_r \bar{\boldsymbol{\lambda}} - \bar{D}^T \mathbf{w} + F_1^T \Theta_r \frac{\mathrm{d}}{\mathrm{d}t}\bar{\boldsymbol{\lambda}}\right). \tag{7.47}$$

Proof. This follows from Theorem 7.7 and the computations before. $\qquad\square$

Theorem 7.14. *Let $\mathbf{u} \in \Upsilon$ be an optimal solution of (7.22)–(7.27) and let $\bar{\mathbf{v}}, \mathbf{z}$ and $\mathbf{c}$ be the corresponding solutions of the systems (7.23)–(7.25) and (7.26)–(7.27) respectively. Then there exist Lagrange multipliers $\bar{\boldsymbol{\lambda}}$ and $\mathbf{d}$ such that $(\bar{\mathbf{v}}, \mathbf{c}, \mathbf{u}, \bar{\boldsymbol{\lambda}}, \mathbf{d})$ satisfy the optimality conditions*

$$\Theta_r^T M \Theta_r \frac{\mathrm{d}}{\mathrm{d}t}\bar{\mathbf{v}}(t) - \Theta_r^T A \Theta_r \bar{\mathbf{v}}(t) - \Theta_r^T \bar{B}\mathbf{u}(t) - \Theta_r^T F_1 \frac{\mathrm{d}}{\mathrm{d}t}\mathbf{u}(t) = \mathbf{0}, \tag{7.48}$$

$$\Theta_r^T M \Theta_r \bar{\mathbf{v}}(0) - \Theta_r^T \mathbf{v}_0 = \mathbf{0}, \tag{7.49}$$

$$\mathbf{z}(t) - \bar{C}\Theta_r \bar{\mathbf{v}}(t) - \bar{D}\mathbf{u}(t) = \mathbf{0}, \tag{7.50}$$

$$M_{ad}\frac{\mathrm{d}}{\mathrm{d}t}\mathbf{c}(t) - A_{ad}(\mathbf{z}(t))\mathbf{c}(t) = \mathbf{0}, \tag{7.51}$$

$$\mathbf{c}(0) = \mathbf{c}_0, \tag{7.52}$$

$$-M_{ad}\frac{\mathrm{d}}{\mathrm{d}t}\mathbf{d} - (A_{ad}(\mathbf{z}))^T \mathbf{d} = \mathbf{0}, \tag{7.53}$$

$$\mathbf{d}(T) + (\mathbf{c}(T) - \mathbf{c}^{foc}) = \mathbf{0}, \tag{7.54}$$

$$\mathbf{w} - \mathcal{F}_{\mathbf{z},\mathbf{c}}^* \mathbf{d} = \mathbf{0}, \tag{7.55}$$

$$-\Theta_r^T M \Theta_r \frac{\mathrm{d}}{\mathrm{d}t}\bar{\boldsymbol{\lambda}} - \Theta_r^T A^T \Theta_r \bar{\boldsymbol{\lambda}} - \Theta_r^T \bar{C}^T \mathbf{w} = \mathbf{0}, \tag{7.56}$$

$$\bar{\boldsymbol{\lambda}}(T) = \mathbf{0}, \tag{7.57}$$

$$\mathcal{P}_{\Upsilon}\left(\sigma \mathbf{u} - \bar{B}^T \Theta_r \bar{\boldsymbol{\lambda}} - \bar{D}^T \mathbf{w} + F_1^T \Theta_r \frac{\mathrm{d}}{\mathrm{d}t}\bar{\boldsymbol{\lambda}}\right) = \mathbf{0}. \tag{7.58}$$

Proof. This follows from Theorem 7.6 and the computations before. $\qquad\square$

7.5 First-order optimality conditions for the original problem

In this subsection we present the first-order optimality conditions for the original optimal control problem (6.51). We can not show the existence of Lagrange multipliers directly for this problem, but we will show a specific sort of equivalence of the optimality conditions for the auxiliary and the original problem and hence, indirectly the existence of Lagrange multipliers. At the beginning we just use the formal technique.

We begin with the original optimization problem

$$\text{Minimize } \frac{1}{2}\|\mathbf{c}(T) - \mathbf{c}^{\text{foc}}\|_{M_{ad}}^2 + \frac{\sigma}{2}\|\mathbf{u}\|_{L^2}^2 \tag{7.59}$$

such that

$$M\frac{\mathrm{d}}{\mathrm{d}t}\mathbf{v} - A\mathbf{v} - B^T\mathbf{p} - F_0\mathbf{u} - F_1\frac{\mathrm{d}}{\mathrm{d}t}\mathbf{u} = 0, \tag{7.60}$$

$$B\mathbf{v} + L\mathbf{u} = 0, \tag{7.61}$$

$$\mathbf{z} - C_1\mathbf{v} = 0, \tag{7.62}$$

$$M\mathbf{v}(0) - \mathbf{v}_0 = 0, \tag{7.63}$$

$$M_{ad}\frac{\mathrm{d}}{\mathrm{d}t}\mathbf{c} - A_{ad}(\mathbf{z})\mathbf{c} = 0, \tag{7.64}$$

$$M_{ad}\mathbf{c}(0) - \mathbf{c}_0 = 0, \tag{7.65}$$

holds true. The Lagrangian $\mathcal{L} : \Upsilon \times \hat{\mathcal{Z}} \to \mathbb{R}$ with

$$\hat{\mathcal{Z}} = \mathcal{Z} \times L^2(0,T;\mathbb{R}^{n_v}) \times L^2(0,T;\mathbb{R}^{n_p}) \times \mathbb{R}^{n_v} \times L^2(0,T;\mathbb{R}^{n_v}) \times L^2(0,T;\mathbb{R}^{n_c}) \times \mathbb{R}^{n_c} \text{ where}$$
$$\mathcal{Z} = H^1(0,T;\mathbb{R}^{n_v}) \times L^2(0,T;\mathbb{R}^{n_p}) \times L^2(0,T;\mathbb{R}^{n_{v,1}}) \times H^1(0,T;\mathbb{R}^{n_c}),$$

reads

$$\mathcal{L}(\mathbf{u}, \mathbf{v}, \mathbf{p}, \mathbf{z}, \mathbf{c}, \boldsymbol{\lambda}, \boldsymbol{\mu}, \boldsymbol{\lambda}_0, \mathbf{w}, \mathbf{d}, \mathbf{d}_0)$$
$$= \frac{1}{2}\|\mathbf{c}(T) - \mathbf{c}^{\text{foc}}\|_{M_{ad}}^2 + \frac{\sigma}{2}\|\mathbf{u}\|_{L^2}^2$$
$$+ \int_0^T (M\frac{\mathrm{d}}{\mathrm{d}t}\mathbf{v} - A\mathbf{v} - B^T\mathbf{p} - F_0\mathbf{u} - F_1\frac{\mathrm{d}}{\mathrm{d}t}\mathbf{u}, \boldsymbol{\lambda})_2 \, \mathrm{d}t$$
$$- \int_0^T (B\mathbf{v} + L\mathbf{u}, \boldsymbol{\mu})_2 \, \mathrm{d}t + \int_0^T (\mathbf{z} - C_1\mathbf{v}, \mathbf{w})_2 \, \mathrm{d}t$$
$$+ (M\mathbf{v}(0) - \mathbf{v}_0, \boldsymbol{\lambda}_0)_2 + \int_0^T (M_{ad}\frac{\mathrm{d}}{\mathrm{d}t}\mathbf{c} - A_{ad}(\mathbf{z})\mathbf{c}, \mathbf{d})_2 \, \mathrm{d}t$$
$$+ (M_{ad}\mathbf{c}(0) - \mathbf{c}_0, \mathbf{d}_0)_2.$$

Determining the derivative of $\mathcal{L}$ with respect to $\mathbf{v}$ in direction $\mathbf{h}$ and setting it to zero, we obtain

$$D_{\mathbf{v}}\mathcal{L}(\mathbf{u}, \mathbf{v}, \mathbf{p}, \mathbf{z}, \mathbf{c}, \boldsymbol{\lambda}, \boldsymbol{\mu}, \boldsymbol{\lambda_0}, \mathbf{w}, \mathbf{d}, \mathbf{d_0})\mathbf{h} = \int_0^T (M\frac{\mathrm{d}}{\mathrm{d}t}\mathbf{h} - A\mathbf{h}, \boldsymbol{\lambda})_2 \, \mathrm{d}t - \int_0^T (B\mathbf{h}, \boldsymbol{\mu})_2 \, \mathrm{d}t$$
$$- \int_0^T (C_1\mathbf{h}, \mathbf{w})_2 \, \mathrm{d}t + (M\mathbf{h}(0), \boldsymbol{\lambda_0})_2 \tag{7.66}$$
$$= 0.$$

Again we can show that the Lagrange multiplier is weakly differentiable like we did for (7.34). Integration by parts yields

$$\int_0^T (M\frac{\mathrm{d}}{\mathrm{d}t}\mathbf{h}, \boldsymbol{\lambda})_2 \, \mathrm{d}t = -\int_0^T (\mathbf{h}, M\frac{\mathrm{d}}{\mathrm{d}t}\boldsymbol{\lambda})_2 \, \mathrm{d}t + (\mathbf{h}(T), M\boldsymbol{\lambda}(T))_2 - (\mathbf{h}(0), M\boldsymbol{\lambda}(0))_2.$$

Inserting this in (7.66) we get

$$-M\frac{\mathrm{d}}{\mathrm{d}t}\boldsymbol{\lambda} - A^T\boldsymbol{\lambda} - B^T\boldsymbol{\mu} - C_1^T\mathbf{w} = \mathbf{0}, \tag{7.67}$$
$$\boldsymbol{\lambda}(T) = \mathbf{0}, \tag{7.68}$$
$$\boldsymbol{\lambda}(0) = \boldsymbol{\lambda_0}. \tag{7.69}$$

Next we consider the derivative of $\mathcal{L}$ with respect to $\mathbf{p}$ in direction $\mathbf{h}$ and set it to be zero. We have

$$D_{\mathbf{p}}\mathcal{L}(\mathbf{u}, \mathbf{v}, \mathbf{p}, \mathbf{z}, \mathbf{c}, \boldsymbol{\lambda}, \boldsymbol{\mu}, \boldsymbol{\lambda_0}, \mathbf{w}, \mathbf{d}, \mathbf{d_0})\mathbf{h} = \int_0^T (-B^T\mathbf{h}, \boldsymbol{\lambda})_2 \, \mathrm{d}t = 0,$$

which leads to the algebraic condition

$$B\boldsymbol{\lambda} = 0. \tag{7.70}$$

Analogously to Section 7.4, we obtain the following conditions

$$-M_{ad}\frac{\mathrm{d}}{\mathrm{d}t}\mathbf{d} - (A_{ad}(\mathbf{z}))^T\mathbf{d} = 0,$$
$$\mathbf{d}(T) = -(\mathbf{c}(T) - \mathbf{c}^{\mathrm{foc}}),$$
$$\mathbf{d}(0) = \mathbf{d_0},$$
$$\mathbf{w} = \mathcal{F}_{\mathbf{z},\mathbf{c}}^*\mathbf{d},$$

for the Lagrange multipliers $\mathbf{d}, \mathbf{d_0}, \mathbf{w}$.

Finally, we compute the derivative of $\mathcal{L}$ with respect to $\mathbf{u}$ in direction $\mathbf{h}$. Setting it to zero, we obtain

$$D_{\mathbf{u}}\mathcal{L}(\mathbf{u}, \mathbf{v}, \mathbf{p}, \mathbf{z}, \mathbf{c}, \boldsymbol{\lambda}, \boldsymbol{\mu}, \boldsymbol{\lambda_0}, \mathbf{w}, \mathbf{d}, \mathbf{d_0})\mathbf{h} = \sigma \int_0^T (\mathbf{u}, \mathbf{h})_2 \, \mathrm{d}t - \int_0^T (L\mathbf{h}, \boldsymbol{\mu})_2 \, \mathrm{d}t - \int_0^T (F_0\mathbf{h}, \boldsymbol{\lambda})_2 \, \mathrm{d}t$$
$$- \int_0^T (F_1\frac{\mathrm{d}}{\mathrm{d}t}\mathbf{h}, \boldsymbol{\lambda})_2 \, \mathrm{d}t = 0. \tag{7.71}$$

Integration by parts yields

$$\int_0^T (F_1 \frac{\mathrm{d}}{\mathrm{d}t} \mathbf{h}, \boldsymbol{\lambda})_2 \, \mathrm{d}t = - \int_0^T (\mathbf{h}, F_1^T \frac{\mathrm{d}}{\mathrm{d}t} \boldsymbol{\lambda})_2 \, \mathrm{d}t + (F_1 \frac{\mathrm{d}}{\mathrm{d}t} \mathbf{h}(T), \boldsymbol{\lambda}(T))_2 - (F_1 \frac{\mathrm{d}}{\mathrm{d}t} \mathbf{h}(0), \boldsymbol{\lambda}(0))_2.$$

Since $\mathbf{h} \in \Upsilon$, $\mathbf{h}(0) = 0$ and due to $\boldsymbol{\lambda}(T) = 0$ we get the condition

$$\mathcal{P}_\Upsilon(\sigma \mathbf{u} - F_0^T \boldsymbol{\lambda} - L^T \boldsymbol{\mu} + F_1^T \frac{\mathrm{d}}{\mathrm{d}t} \boldsymbol{\lambda}) = 0.$$

Theorem 7.15. *Let* $\mathbf{u} \in \Upsilon$ *be an arbitrary vector. Then*

$$\nabla J(\mathbf{h}) = \mathcal{P}_\Upsilon(\sigma \mathbf{u} - F_0^T \boldsymbol{\lambda} - L^T \boldsymbol{\mu} + F_1^T \frac{\mathrm{d}}{\mathrm{d}t} \boldsymbol{\lambda}). \tag{7.72}$$

Theorem 7.16. *Let* $\mathbf{u} \in \Upsilon$ *be an optimal solution of* (7.22)–(7.27) *and let* $\mathbf{v}, \mathbf{p}, \mathbf{z}$ *and* $\mathbf{c}$ *be the corresponding solutions of the systems* (7.60)–(7.63) *and* (7.64)–(7.65) *respectively. Then there exist Lagrange multipliers* $\boldsymbol{\lambda}, \boldsymbol{\mu}$ *and* $\mathbf{d}$ *such that* $(\mathbf{v}, \mathbf{p}, \mathbf{c}, \mathbf{u}, \boldsymbol{\lambda}, \boldsymbol{\mu}, \mathbf{d})$ *satisfy the optimality conditions*

$$M \frac{\mathrm{d}}{\mathrm{d}t} \mathbf{v}(t) - A\mathbf{v}(t) - B^T \mathbf{p}(t) - F_0 \mathbf{u}(t) - F_1 \frac{\mathrm{d}}{\mathrm{d}t} \mathbf{u}(t) = \mathbf{0}, \tag{7.73}$$

$$B\mathbf{v}(t) + L\mathbf{u}(t) = \mathbf{0}, \tag{7.74}$$

$$\mathbf{z}(t) - C_1 \mathbf{v}(t) = \mathbf{0}, \tag{7.75}$$

$$M\mathbf{v}(0) - \mathbf{v}_0 = \mathbf{0}, \tag{7.76}$$

$$M_{ad} \frac{\mathrm{d}}{\mathrm{d}t} \mathbf{c}(t) - A_{ad}(\mathbf{z}(t))\mathbf{c}(t) = \mathbf{0}, \tag{7.77}$$

$$M_{ad}\mathbf{c}(0) = \mathbf{c}_0, \tag{7.78}$$

$$-M_{ad} \frac{\mathrm{d}}{\mathrm{d}t} \mathbf{d}(t) - (A_{ad}(\mathbf{z}(\mathbf{t})))^T \mathbf{d}(t) = \mathbf{0}, \tag{7.79}$$

$$\mathbf{d}(T) + (\mathbf{c}(T) - \mathbf{c}^{foc}) = \mathbf{0}, \tag{7.80}$$

$$\mathbf{w} = \mathcal{F}_{\mathbf{z},\mathbf{c}}^* \mathbf{d} = \mathbf{0}, \tag{7.81}$$

$$-M \frac{\mathrm{d}}{\mathrm{d}t} \boldsymbol{\lambda}(t) - A^T \boldsymbol{\lambda}(t) - B^T \boldsymbol{\mu}(t) - C_1^T \mathbf{w}(t) = \mathbf{0}, \tag{7.82}$$

$$B\boldsymbol{\lambda}(t) = \mathbf{0}, \tag{7.83}$$

$$M\boldsymbol{\lambda}(T) = \mathbf{0}, \tag{7.84}$$

$$\mathcal{P}_\Upsilon(\sigma \mathbf{u} - F_0^T \boldsymbol{\lambda} - L^T \boldsymbol{\mu} + F_1^T \frac{\mathrm{d}}{\mathrm{d}t} \boldsymbol{\lambda}) = \mathbf{0}. \tag{7.85}$$

Proof. We investigate the relation of the first-order optimality conditions (7.48)–(7.58) and (7.73)–(7.85) of the optimal control problems (7.22)–(7.27) and (7.59)–(7.65), respectively. It turns out that the optimality conditions of both problems are equivalent. Thus, we show directly the existence of Lagrange multipliers for the original problem.

Since the considered equations are of exactly the same form as treated by Theorem 6.4, we can skip most of the justifications and just apply the transformations described in the proof of that theorem to the situation at hand.

If we start with the equations

$$-\Theta_r^T M \Theta_r \frac{\mathrm{d}}{\mathrm{d}t}\bar{\boldsymbol{\lambda}} - \Theta_r^T A^T \Theta_r \bar{\boldsymbol{\lambda}} - \Theta_r^T \bar{C}^T \mathbf{w} = 0,$$

$$\bar{\boldsymbol{\lambda}}(T) = 0,$$

and we define $\boldsymbol{\lambda} := \Theta_r \bar{\boldsymbol{\lambda}}$, then we obtain by using this definition the system

$$-\Theta_r^T M \frac{\mathrm{d}}{\mathrm{d}t}\boldsymbol{\lambda} - \Theta_r^T A^T \boldsymbol{\lambda} - \Theta_r^T \bar{C}^T \mathbf{w} = 0, \qquad (7.86)$$

$$\Theta_r^T M \boldsymbol{\lambda}(T) = 0. \qquad (7.87)$$

Note that for $\Pi = \Theta_l \Theta_r^T$, the identity

$$\Pi^T \boldsymbol{\lambda} = \Theta_r \Theta_l^T \Theta_r \bar{\boldsymbol{\lambda}} = \Theta_r \bar{\boldsymbol{\lambda}} = \boldsymbol{\lambda}$$

holds true. We multiply equations (7.86)–(7.87) from the left by Θ_l and obtain the equivalent system

$$-\Pi M \Pi^T \frac{\mathrm{d}}{\mathrm{d}t}\boldsymbol{\lambda} - \Pi A^T \Pi^T \boldsymbol{\lambda} - \Pi C_1^T \mathbf{w} = \mathbf{0}, \qquad (7.88)$$

$$\Pi M \Pi^T \boldsymbol{\lambda}(T) = \mathbf{0}, \qquad (7.89)$$

which possesses a differential algebraic structure.

Let now an additional multiplier $\boldsymbol{\mu}$ be defined as

$$\boldsymbol{\mu} = -(BM^{-1}B^T)^{-1}BM^{-1}(A^T \boldsymbol{\lambda} + C_1^T \mathbf{w}). \qquad (7.90)$$

Then, system (7.88)–(7.89) together with (7.90) is equivalent to the system

$$-M \frac{\mathrm{d}}{\mathrm{d}t}\boldsymbol{\lambda} - A^T \boldsymbol{\lambda} - B^T \boldsymbol{\mu} - C_1^T \mathbf{w} = \mathbf{0}, \qquad (7.91)$$

$$B\boldsymbol{\lambda} = \mathbf{0}, \qquad (7.92)$$

$$M\boldsymbol{\lambda}(T) = \mathbf{0}. \qquad (7.93)$$

$$\qquad (7.94)$$

This is the case, because using $\boldsymbol{\lambda} = \Pi^T \boldsymbol{\lambda} + (\boldsymbol{\lambda} - \Pi^T \boldsymbol{\lambda}) = \boldsymbol{\lambda}_1 + \boldsymbol{\lambda}_2$ leads to

$$\boldsymbol{\lambda}_2 = M^{-1}B^T(BM^{-1}B^T)^{-1}B\boldsymbol{\lambda} = \mathbf{0},$$

due to the algebraic condition $B\boldsymbol{\lambda} = \mathbf{0}$.

In the expression (7.47) of the gradient of the auxiliary system the term $-\bar{B}^T \Theta_r \bar{\boldsymbol{\lambda}} - \bar{D}^T \mathbf{w}$ appears, where the matrices $\bar{B}, \bar{D}$ are defined as

$$\bar{B} = F_0 - AM^{-1}B^T(BM^{-1}B^T)^{-1}L,$$
$$\bar{D} = -C_1 M^{-1}B^T(BM^{-1}B^T)^{-1}L.$$

Then we have

$$
\begin{aligned}
-\bar{B}^T \Theta_r \bar{\boldsymbol{\lambda}} - \bar{D}^T \mathbf{w} &= -(F_0^T - L^T (BM^{-1}B^T)^{-T}BM^{-1}A^T)\Theta_r \bar{\boldsymbol{\lambda}} \\
&\quad + L^T (BM^{-1}B^T)^{-1}BM^{-1}C_1^T \mathbf{w} \\
&= -F_0^T \boldsymbol{\lambda} + L^T (BM^{-1}B^T)^{-T}BM^{-1}(A^T \boldsymbol{\lambda} + C_1^T \mathbf{w}) \\
&= -F_0^T \boldsymbol{\lambda} - L^T \boldsymbol{\mu}.
\end{aligned}
$$

Taking into account that $F_1^T \Theta_r \frac{\mathrm{d}}{\mathrm{d}t}\bar{\boldsymbol{\lambda}} = F_1 \frac{\mathrm{d}}{\mathrm{d}t}\boldsymbol{\lambda}$, we obtain that (7.47) is equivalent to (7.72). The Lagrange multiplier $\mathbf{d}$ of the original system can be chosen the same as for the auxiliary system.

As a result, we have shown the existence of the Lagrange multipliers such that the following first-order optimality conditions

$$
\begin{aligned}
-M \frac{\mathrm{d}}{\mathrm{d}t}\boldsymbol{\lambda}(t) - A\boldsymbol{\lambda}(t) - B^T \boldsymbol{\mu}(t) - C^T \mathbf{w}(t) &= \mathbf{0}, \\
B\boldsymbol{\lambda}(t) &= \mathbf{0}, \\
M\boldsymbol{\lambda}(T) &= \mathbf{0}, \\
-M_{ad} \frac{\mathrm{d}}{\mathrm{d}t}\mathbf{d} - (A_{ad}(\mathbf{z}))^T \mathbf{d} &= \mathbf{0}, \\
\mathbf{d}(T) + (\mathbf{c}(T) - \mathbf{c}^{\mathrm{foc}}) &= \mathbf{0}, \\
\mathbf{w} - \mathcal{F}_{\mathbf{z},\mathbf{c}}^* \mathbf{d} &= \mathbf{0}, \\
\mathcal{P}_\Upsilon(\sigma \mathbf{u} - F_0^T \boldsymbol{\lambda} - L^T \boldsymbol{\mu} + F_1^T \frac{\mathrm{d}}{\mathrm{d}t}\boldsymbol{\lambda}) &= \mathbf{0}.
\end{aligned}
$$

hold true. □

Due to this result, the computation of the auxiliary system and especially the projector matrices Θ_r, Θ_l and the projector Π is not needed, and we have a justification for the use of the formal Lagrange technique.

Proof of Theorem 7.15. This follows from Theorem 7.7 and the investigations from the proof of Theorem 7.16. □

7.6 The gradient method

In this section we finally apply the gradient descent method [4] in order to solve the optimal control problem (7.59)–(7.65).

The idea of this method is to take steps proportional to the negative gradient of the main functional in order to find a local minimum. There exist algorithms to determine the length of these steps. In our case, we use the Armijo line search. [24]

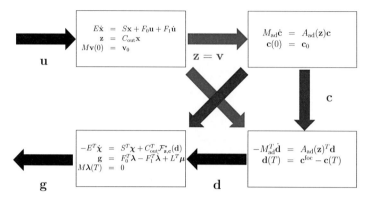

Figure 7.1: The coupled system of equations.

If we want to apply the gradient method, the gradient of the functional we want to minimize is required in the algorithm. Figure 7.1 shows the four equations which have to be solved in each step of the optimization process in order to compute the current gradient. We chose $F_0^T \lambda + L^T \mu - F_1^T \frac{d}{dt} \lambda$ as the output of the adjoint equation, because this expression is part of the description of the gradient.

Algorithm 1 summarizes the gradient method applied to the optimization problem (7.59)–(7.65). Note that it is very expensive to solve those four equations in each step of the optimization process. This problem leads to the idea of using model order reduction techniques. The solvability of the equations (7.73)–(7.84) follows from the following theorem:

Theorem 7.17. *Let $\mathbf{u}$ be an arbitrarily chosen control from Υ. Then, the equations (7.73)–(7.84) have a unique solution.*

Proof. This follows immediately from Theorems 6.5, 6.9 and 7.7. $\qquad \square$

If we can find reduced order models for the state equations (7.73)–(7.76), (7.77)–(7.78) and the adjoint equations (7.79)–(7.81), (7.82)–(7.84), computational time in the gradient method could be decreased. In the next chapter we will discuss appropriate techniques for computing reduced-order models for the linear Stokes–Brinkman and adjoint equations and the nonlinear advection-diffusion and adjoint equations, respectively.

7.7 Error estimation

Next, we establish error estimates for the optimality system. There already exist a posteriori error estimates developed for linear quadratic optimal control problems, in other words for

Algorithm 1 Gradient method

1: **Input:** $\mathbf{u}_{\text{start}}, \epsilon_{\text{tol}}$
2: **Output:** $\mathbf{u}^*$
3: Initialization: $\mathbf{u}^{(0)} = \mathbf{u}_{\text{start}}, k = 0$
4: **while** $\|\mathbf{g}\| > \epsilon_{\text{tol}}$ **do**
5:      Solve the discretized Stokes–Brinkman equation (7.73)–(7.76) for $\mathbf{z}$.
6:      Solve the discretized advection-diffusion equation (7.77)–(7.78) for $\mathbf{c}$.
7:      Solve the adjoint advection-diffusion equation (7.79)–(7.81) for $\mathbf{d}, \mathbf{w}$.
8:      Solve the adjoint Stokes–Brinkman equation (7.82)–(7.84) for $\boldsymbol{\lambda}, \boldsymbol{\mu}$.
9:      Compute the gradient $\mathbf{g}$ from (7.85).
10:     Compute a step size α using Armijo line search
11:     $\mathbf{u}^{(k+1)} = \mathbf{u}^{(k)} - \alpha \mathbf{g}$
12:     $k = k + 1$
13: **end while**

tracking type functionals governed by linear equations, see [44]. Furthermore, in [48], also estimates for nonlinear equations were developed. We follow [48] for notation and the idea of the error estimate. The idea of the a posteriori estimate is to make use of the necessary optimality conditions which come up from the optimal control problem. The twice Fréchet differentiability of $\mathcal{S}_1$ and $\mathcal{S}_2$ will be used in this section to derive an error estimate using the approach from [44]. Note that the idea can also be applied to the reduced system from the next chapter. We denote by $\mathbf{u}^* \in \Upsilon$ the optimal control of the original problem (7.59) and by $\bar{\mathbf{u}} \in \Upsilon$ some arbitrary control. Then the two variational inequalities

$$(\nabla J(\bar{\mathbf{u}}) + \boldsymbol{\zeta}, \mathbf{u}^* - \bar{\mathbf{u}})_{L^2} \geq 0 \tag{7.95}$$
$$(\nabla J(\mathbf{u}^*), \bar{\mathbf{u}} - \mathbf{u}^*)_{L^2} \geq 0 \tag{7.96}$$

hold true for some function $\boldsymbol{\zeta}$, where J is the main functional. The second inequality (7.96) follows from the necessary optimality conditions from Section 7. A possible choice of $\boldsymbol{\zeta}$ in (7.95) is

$$\boldsymbol{\zeta}(t) := -\nabla J(\bar{\mathbf{u}})(t),$$

because then the inequality (7.96) is fulfilled trivially. Adding both inequalities, we obtain

$$
\begin{aligned}
0 &\leq (\nabla J(\bar{\mathbf{u}}) + \boldsymbol{\zeta}, \mathbf{u}^* - \bar{\mathbf{u}})_{L^2} + (\nabla J(\mathbf{u}^*), \bar{\mathbf{u}} - \mathbf{u}^*)_{L^2} \\
&= (\nabla J(\bar{\mathbf{u}}) + \boldsymbol{\zeta}, \mathbf{u}^* - \bar{\mathbf{u}})_{L^2} - (\nabla J(\mathbf{u}^*), \mathbf{u}^* - \bar{\mathbf{u}})_{L^2} \\
&= (\nabla J(\bar{\mathbf{u}}) - \nabla J(\mathbf{u}^*), \mathbf{u}^* - \bar{\mathbf{u}})_{L^2} + (\boldsymbol{\zeta}, \mathbf{u}^* - \bar{\mathbf{u}})_{L^2}.
\end{aligned}
$$

The mean value theorem leads to

$$-D^2 J(\hat{\mathbf{u}})(\mathbf{u}^* - \bar{\mathbf{u}}, \mathbf{u}^* - \bar{\mathbf{u}}) + (\boldsymbol{\zeta}, \mathbf{u}^* - \bar{\mathbf{u}})_{L^2} \geq 0, \tag{7.97}$$

where $\hat{\mathbf{u}} \in \{\mathbf{v} \in \Upsilon \mid \mathbf{v} = s\mathbf{u}^* + (1-s)\bar{\mathbf{u}}, \quad s \in [0,1]\}$.

Assume that J is strongly convex on an open neighbourhood of $\mathbf{u}^*$, that is, there exists a constant $\alpha > 0$ and a constant $\delta > 0$ such that for all $\mathbf{u}$ in a neighbourhood of $\mathbf{u}^*$ it holds

$$D^2 J(\mathbf{u})(\mathbf{h}, \mathbf{h}) \geq \alpha \|\mathbf{h}\|_{L^2}^2 \tag{7.98}$$

for $\|\mathbf{u}^* - \mathbf{u}\|_{L^2} \leq \delta$. Then it follows from (7.97) and (7.98) that

$$\alpha \|\mathbf{u}^* - \bar{\mathbf{u}}\|_{L^2}^2 \leq D^2 J(\hat{\mathbf{u}})(\mathbf{u}^* - \bar{\mathbf{u}}, \mathbf{u}^* - \bar{\mathbf{u}}) \leq (\boldsymbol{\zeta}, \mathbf{u}^* - \bar{\mathbf{u}})_{L^2} \leq \|\boldsymbol{\zeta}\|_{L^2} \|\mathbf{u}^* - \bar{\mathbf{u}}\|_{L^2},$$

and, therefore,

$$\|\mathbf{u}^* - \bar{\mathbf{u}}\|_{L^2} \leq \frac{1}{\alpha} \|\boldsymbol{\zeta}\|_{L^2}. \tag{7.99}$$

Note that (7.99) is only a local estimate, because it relies on the local convexity of J. If J is strongly convex overall, then the estimate will work globally.

Remark 7.18. Without further restrictions of the control we considered an open control space in this work, such that it holds

$$DJ(\mathbf{u}^*) = 0,$$

for the optimal control. If restrictions like $a \leq u^{(i)} \leq b, i = 1, 2$ for some $a, b \in \mathbb{R}$ for the control are added, the gradient in the minimum is not necessarily zero any longer. In this case, the definition of $\boldsymbol{\zeta}$ has to be modified slightly, see [44] for details.

8 Model order reduction

In this chapter, we discuss several model order reduction techniques. The first part of this chapter deals with reduction methods for linear time-invariant equations using the iterative rational Krylov algorithm as well as the balanced truncation approach. The application of these methods to the semi-discretized Stokes–Brinkman equation (5.20)–(5.22) is also discussed. In the second part of the chapter, we focus on nonlinear model reduction. Proper orthogonal decomposition (POD) and the discrete empirical interpolation method (DEIM) are described. We also present an algorithm for matrix DEIM which in combination with POD is used for model reduction of the semi-discretized advection-diffusion equation (5.23)–(5.24).

8.1 Model reduction by tangential interpolation

In this section, we briefly review a tangential interpolation model reduction approach from [38, 54] and present the iterative rational Krylov algorithm which is an iterative method to compute a reduced-order model. We also discuss the application of this algorithm to the Stokes–Brinkman equation (5.20)–(5.22). The basic idea to compute a reduced-order model consists in minimizing an error functional. This was an often investigated problem, see [45, 42, 73]. As finding global minimizers was difficult, the approach was to fulfill the first-order necessary optimality conditions of the minimization problem [54]. For systems with multiple inputs and multiple outputs – referred to as the MIMO systems – the existence of a global minimizer is still an open problem. In [36] the Iterative Rational Krylov algorithm (IRKA) was investigated, an iterative method which computes a solution such that the first-order necessary optimality conditions of a certain optimization problem are fulfilled. The convergence of IRKA was examined in [28].

For a description of the method we start with the linear time-invariant system

$$E\frac{\mathrm{d}}{\mathrm{d}t}\mathbf{x}(t) = F\mathbf{x}(t) + G\mathbf{u}(t), \quad E\mathbf{x}(0) = \mathbf{x}_0, \tag{8.1}$$

$$\mathbf{y}(t) = C\mathbf{x}(t), \tag{8.2}$$

where $E, F \in \mathbb{R}^{n \times n}, G \in \mathbb{R}^{n \times m}, C \in \mathbb{R}^{q \times n}$ and E is assumed to be nonsingular. Furthermore, $\mathbf{u}(t) \in \mathbb{R}^m$ denotes the input, $\mathbf{x}(t) \in \mathbb{R}^n$ the state and $\mathbf{y}(t) \in \mathbb{R}^q$ the output of the system (8.1)–(8.2).

The transfer function $H : \mathbb{C} \to \mathbb{C}^{q \times m}$ of this system is defined as

$$H(s) = C(sE - F)^{-1}G.$$

Let $\mathcal{H}_2$ denote the space of stable, proper transfer functions. Then the $\mathcal{H}_2$-norm of the transfer function H is given as

$$\|H\|_{\mathcal{H}_2} = \left(\frac{1}{2\pi} \int_{-\infty}^{\infty} \|H(i\omega)\|_F^2 \, \mathrm{d}\omega \right)^{\frac{1}{2}},$$

where $\| \cdot \|_F$ denotes the Frobenius matrix norm.

The aim is to find a reduced-order model

$$\hat{E}\frac{\mathrm{d}}{\mathrm{d}t}\hat{\mathbf{x}}(t) = \hat{F}\hat{\mathbf{x}}(t) + \hat{G}\mathbf{u}(t), \quad \hat{E}\hat{\mathbf{x}}(0) = \hat{\mathbf{x}}_0, \tag{8.3}$$

$$\hat{\mathbf{y}}(t) = \hat{C}\hat{\mathbf{x}}(t), \tag{8.4}$$

where $\hat{E}, \hat{F} \in \mathbb{R}^{\ell \times \ell}, \hat{G} \in \mathbb{R}^{\ell \times m}, \hat{C} \in \mathbb{R}^{q \times \ell}$ and $\ell \ll n$. The transfer function of the reduced-order model is then given by

$$\hat{H}(s) = \hat{C}(s\hat{E} - \hat{F})^{-1}\hat{G}.$$

The reduced-order model (8.3)–(8.4) can be obtained by the Petrov-Galerkin projection

$$W^T EV \frac{\mathrm{d}}{\mathrm{d}t}\hat{\mathbf{x}}(t) = W^T FV\hat{\mathbf{x}}(t) + W^T G\mathbf{u}(t), \quad W^T EV\hat{\mathbf{x}}(0) = W^T \mathbf{x}_0, \tag{8.5}$$

$$\hat{\mathbf{y}}(t) = CV\hat{\mathbf{x}}(t), \tag{8.6}$$

for some projection matrices $V, W \in \mathbb{R}^{n \times \ell}$. The goal of model reduction techniques is to find these matrices such that the error $\hat{\mathbf{y}} - \mathbf{y}$ is small with respect to some norm, or, equivalently, the transfer function $\hat{H}$ approximates the original transfer function H well.

The idea of model reduction via tangential interpolation is to compute V and W in such a way that the transfer function $\hat{H}$ of the approximate system (8.3)–(8.4) interpolates the original transfer function H at several points in the complex plane along several directions. This means, that the conditions

$$\mathbf{c}_i^T H(\sigma_i) = \mathbf{c}_i^T \hat{H}(\sigma_i), \quad i = 1, \dots, r \tag{8.7}$$

$$H(\mu_i)\mathbf{b}_i = \hat{H}(\mu_i)\mathbf{b}_i, \quad i = 1, \dots, r, \tag{8.8}$$

are fulfilled, where $\sigma_i, \mu_i \in \mathbb{C}$ are the interpolation points and $\mathbf{c}_i \in \mathbb{C}^q, \mathbf{b}_i \in \mathbb{C}^m$ are the left and right tangential directions, respectively.

The following theorem was shown in [37].

Theorem 8.1. *Let* $\sigma, \mu \in \mathbb{C}$ *be such that* $sE - F$ *and* $s\hat{E} - \hat{F}$ *are both invertible for* $s = \sigma, \mu$, *and let* $\mathbf{b} \in \mathbb{C}^m$ *and* $\mathbf{c} \in \mathbb{C}^q$ *be fixed nontrivial vectors.*

1. If

$$((\sigma E - F)^{-1}E)^{j-1}(\sigma E - F)^{-1}G\mathbf{b} \in \mathrm{Im}(V), \quad j = 1, \dots, N,$$

then $H^{(l)}(\sigma)\mathbf{b} = \hat{H}^{(l)}(\sigma)\mathbf{b}$ *for* $l = 0, 1, \dots, N - 1$.

2. If

$$((\mu E - F)^{-T} E^T)^{j-1} (\mu E - F)^{-T} C^T \mathbf{c} \in \text{Im}(W), \quad j = 1, \ldots, M,$$

then $\mathbf{c}^T H^{(l)}(\mu) = \mathbf{c}^T \hat{H}^{(l)}(\mu)$ *for* $l = 0, 1, \ldots, M - 1$.

3. If both conditions above are fulfilled and if $\sigma = \mu$, *then* $\mathbf{c}^T H^{(l)}(\sigma)\mathbf{b} = \mathbf{c}^T \hat{H}^{(l)}(\sigma)\mathbf{b}$ *for* $l = 0, 1, \ldots, N + M + 1$.

That means, if V and W are constructed in such a way, that the assumptions of Theorem 8.1 are satisfied, then the interpolation conditions (8.7) and (8.8) will be fulfilled. The projection matrices are in this case given by

$$V = [(\sigma_1 E - F)^{-1} G \mathbf{b}_1, \ldots (\sigma_r E - F)^{-1} G \mathbf{b}_r],$$
$$W = [(\mu_1 E - F)^{-T} C^T \mathbf{c}_1, \ldots, (\mu_r E - F)^{-T} C^T \mathbf{c}_r].$$

8.1.1 Iterative rational Krylov algorithm

At this point, it is still an open question how the interpolation points and the tangential directions should be chosen. It was suggested in [38] to determine the projection matrices by solving the following optimization problem

$$\text{minimize} \| H - \hat{H} \|_{\mathcal{H}_2}, \tag{8.9}$$

among all systems of state space dimension ℓ.

This optimization problem is nonconvex which makes it very difficult to find a global minimizer if there even exists one. In practice, one tries to find a local minimizer instead which is supposed to fulfill the first-order necessary optimality conditions.

Originally, there were two different approaches for solving this optimization problem: a Lyapunov-based approach [72, 45] and an interpolation based approach [37, 36]. It was shown in [38] that both approaches are equivalent.

It was shown in [54], that the optimality conditions for (8.9) read

$$H(-\lambda_i)\mathbf{b}_i = \hat{H}(-\lambda_i)\mathbf{b}_i, \tag{8.10}$$
$$\mathbf{c}_i^T H(-\lambda_i) = \mathbf{c}_i^T \hat{H}(-\lambda_i), \tag{8.11}$$
$$\mathbf{c}_i^T H'(-\lambda_i)\mathbf{b}_i = \mathbf{c}_i^T \hat{H}'(-\lambda_i)\mathbf{b}_i, \tag{8.12}$$

where λ_i are the eigenvalues of the pencil $\lambda \hat{E} - \hat{F}$. This leads to the optimal interpolation points, choosing $\sigma_i = \mu_i = -\lambda_i$. But we can not compute them directly because we do not know $\hat{E}$ and $\hat{F}$ a priori.

In [34, 36], an iterative procedure called the *Iterative Rational Krylov Algorithm (IRKA)* was presented. It generates a reduced-order model (8.3)–(8.4) which ensures that the first-order necessary optimality conditions (8.10)–(8.12) are fulfilled. We summarize this procedure in Algorithm 2.

Algorithm 2 IRKA

1: **Input:** An initial shift selection $\sigma_i, i = 1, \ldots, \ell$ and initial tangential directions $\mathbf{b}_1, \ldots, \mathbf{b}_\ell$ and $\mathbf{c}_1, \ldots, \mathbf{c}_\ell$.

2: **Output:** W, V

3: $\quad W = [(\bar{\sigma}_1 E^T - F^T)^{-1} C^T \mathbf{c}_1, \ldots, (\bar{\sigma}_\ell E^T - F^T)^{-1} C^T \mathbf{c}_\ell]$

4: $\quad V = [(\sigma_1 E - F)^{-1} G \mathbf{b}_1, \ldots, (\sigma_\ell E - F)^{-1} G \mathbf{b}_\ell]$

5: **while** not converged **do**

6: $\quad \hat{F} = W^T F V, \hat{E} = W^T E V$

7: $\quad$ Compute Y, Z with $Y^* \hat{F} Z = \text{diag}(\lambda_1, \ldots, \lambda_\ell)$ and $Y^* \hat{E} Z = I_\ell$, where the $\quad$ columns of $Z = [\mathbf{z}_1, \ldots, \mathbf{z}_\ell]$ and $Y = [\mathbf{y}_1, \ldots, \mathbf{y}_\ell]$ are, respectively, the right $\quad$ and left eigenvectors of $\lambda \hat{E} - \hat{F}$.

8: $\quad \sigma_i = -\lambda_i, \mathbf{b}_i^T = \mathbf{y}_i^* \hat{G}$ and $\mathbf{c}_i = \hat{C} \mathbf{z}_i$ for $i = 1, \ldots, \ell$

9: $\quad W = [(\bar{\sigma}_1 E^T - F^T)^{-1} C^T \mathbf{c}_1, \ldots, (\bar{\sigma}_\ell E^T - F^T)^{-1} C^T \mathbf{c}_\ell]$

10: $\quad V = [(\sigma_1 E - F)^{-1} G \mathbf{b}_1, \ldots, (\sigma_\ell E - F)^{-1} G \mathbf{b}_\ell]$

11: **end while**

12: $\hat{F} = W^T F V, \hat{E} = W^T E V, \hat{G} = W^T G, \hat{C} = C V$

8.1.2 IRKA for the Stokes–Brinkman equation

In [39], the tangential interpolation approach was extended to Stokes-type systems. Here, we give a brief review of this method and in the outline, we follow [39].

We consider a Stokes–Brinkman system of the form

$$M \frac{\mathrm{d}}{\mathrm{d}t} \mathbf{v}(t) = A\mathbf{v}(t) + B^T \mathbf{p}(t) + F_0 \mathbf{u}(t) + F_1 \frac{\mathrm{d}}{\mathrm{d}t} \mathbf{u}(t), \quad M\mathbf{v}(0) = \mathbf{v}_0 \quad (8.13)$$

$$0 = B\mathbf{v}(t) + L\mathbf{u}(t), \quad (8.14)$$

$$\mathbf{z}(t) = C_1 \mathbf{v}(t), \quad (8.15)$$

where the state is $\mathbf{x}(t) = \left[\mathbf{v}^T(t), \mathbf{p}^T(t)\right]^T \in \mathbb{R}^n$ with $\mathbf{v}(t) \in \mathbb{R}^{n_v}$, $\mathbf{p}(t) \in \mathbb{R}^{n_p}$ and $n_v + n_p = n$, the input is $\mathbf{u}(t) \in \mathbb{R}^m$, the output is $\mathbf{z}(t) \in \mathbb{R}^q$, and $M, A \in \mathbb{R}^{n_v \times n_v}$, $B^T \in \mathbb{R}^{n_v \times n_p}$, $F_0, F_1 \in \mathbb{R}^{n_v \times m}$, $L \in \mathbb{R}^{n_p \times m}$, $C_1 \in \mathbb{R}^{q \times n_v}$.

We assume that M is symmetric and positive definite, B^T has full column rank. Then $BM^{-1}B^T$ is nonsingular. It has been shown in Section 7.2 that system (8.13)–(8.15) is equivalent to the system

$$\Pi M \Pi^T \frac{\mathrm{d}}{\mathrm{d}t} \mathbf{v}_1(t) = \Pi A \Pi^T \mathbf{v}_1(t) + \Pi \bar{B} \mathbf{u}(t) + \Pi F_1 \frac{\mathrm{d}}{\mathrm{d}t} \mathbf{u}(t), \quad \Pi M \Pi^T \mathbf{v}_1(0) = \Pi \mathbf{v}_0, \quad (8.16)$$

$$\Pi^T \mathbf{v}_1(t) = \mathbf{v}_H(t), \quad (8.17)$$

$$\mathbf{z}(t) = \bar{C} \Pi^T \mathbf{v}_1(t) + \bar{D} \mathbf{u}(t), \quad (8.18)$$

in the sense that they both have the same output $\mathbf{z}$ for the input $\mathbf{u}$. Here $\mathbf{v}_1 = \Pi^T \mathbf{v}$ with $\Pi = I - B^T (BM^{-1}B^T)^{-1}BM^{-1}$ and the matrices $\bar{B}, \bar{C}$ and $\bar{D}$ are given as

$$\bar{B} := F_0 - AM^{-1}B^T(BM^{-1}B^T)^{-1}L,$$
$$\bar{C} := C_1,$$
$$\bar{D} := C_1 M^{-1} B^T (BM^{-1}B^T)^{-1}L.$$

If we want to apply the Algorithm 2 to system (8.16)–(8.18), we need to invert the matrix $(\sigma\Pi M\Pi^T - \Pi A\Pi^T)$ with $\sigma \in \mathbb{C}$ which is singular. To overcome this difficulty, it was proposed in [39] to use a pseudo-inverse $(\sigma\Pi M\Pi^T - \Pi A\Pi^T)^I$ of $(\sigma\Pi M\Pi^T - \Pi A\Pi^T)$ with respect to the projectors Π and Π^T instead of the inverse which does not exist. For a matrix $N \in \mathbb{R}^{n\times n}$, a pseudo-inverse N^I with respect to the projectors Π and Π^T is defined as a unique solution of the matrix equations

$$N^I N N^I = N^I,$$
$$N N^I = \Pi,$$
$$N^I N = \Pi^T.$$

Using the pseudo-inverse, the projection matrices for system (8.16)–(8.18) can be determined as

$$V = [(\sigma_1\Pi M\Pi^T - \Pi A\Pi^T)^I \Pi[\bar{B} \ F_1]\mathbf{b}_1, \ldots, (\sigma_r\Pi M\Pi^T - \Pi A\Pi^T)^I \Pi[\bar{B} \ F_1]\mathbf{b}_r],$$
$$W = [(\sigma_1\Pi M\Pi^T - \Pi A^T\Pi^T)^I \Pi\bar{C}^T\mathbf{c}_1, \ldots, (\sigma_r\Pi M\Pi^T - \Pi A^T\Pi^T)^I \Pi\bar{C}^T\mathbf{c}_r].$$

The columns $\mathbf{v}_i$ and $\mathbf{w}_i$ of these matrices can be obtained by solving the following linear systems

$$\begin{bmatrix} \sigma M - A & B^T \\ B & 0 \end{bmatrix} \begin{bmatrix} \mathbf{v}_i \\ \mathbf{e} \end{bmatrix} = \begin{bmatrix} [\bar{B} \ F_1]\mathbf{b}_i \\ 0 \end{bmatrix}, \tag{8.19}$$

$$\begin{bmatrix} \sigma M - A^T & B^T \\ B & 0 \end{bmatrix} \begin{bmatrix} \mathbf{w}_i \\ \mathbf{q} \end{bmatrix} = \begin{bmatrix} \bar{C}^T\mathbf{c}_i \\ 0 \end{bmatrix}. \tag{8.20}$$

Note that the matrix $\begin{bmatrix} \sigma M - A & B^T \\ B & 0 \end{bmatrix}$ is invertible, if $B(\sigma M - A)^{-1}B^T$ is invertible. Note that the matrix $(\sigma M - A)$ is regular for $\sigma = \sigma_1, \ldots, \sigma_r$, because σ_i are not eigenvalues of the pencil $\lambda M - A$.

Using IRKA for the Stokes–Brinkman equation (8.13)–(8.15) as presented in Algorithm 3, an explicit computation of Π is not required anymore and the reduced-order model can be computed directly using the original matrices. As a result we obtain the reduced-order model

$$\hat{M}\frac{\mathrm{d}}{\mathrm{d}t}\hat{\mathbf{v}}(t) = \hat{A}\hat{\mathbf{v}}(t) + \hat{B}_1\mathbf{u}(t) + \hat{B}_2\frac{\mathrm{d}}{\mathrm{d}t}\mathbf{u}(t),$$
$$\hat{\mathbf{z}}(t) = \hat{C}\hat{\mathbf{v}}(t) + \hat{D}\mathbf{u}(t),$$

where $\hat{\mathbf{z}}(t)$ approximates the original output $\mathbf{z}(t)$ of system (8.13)–(8.15), $\hat{B} = [\hat{B}_1 \ \hat{B}_2]$ and $\hat{D} = \bar{D}$.

Algorithm 3 IRKA for the Stokes–Brinkman equation

1: **Input:** An initial shift selection $\sigma_i, i = 1, \ldots, \ell$ and initial tangential directions $\mathbf{b}_1, \ldots, \mathbf{b}_\ell$ and $\mathbf{c}_1, \ldots, \mathbf{c}_\ell$.

2: **Output:** W, V

3: **for** $i = 1, \ldots, \ell$ **do**

4:    Solve

$$\begin{bmatrix} \sigma_i M - A & B^T \\ B & 0 \end{bmatrix} \begin{bmatrix} \mathbf{v}_i \\ \mathbf{z} \end{bmatrix} = \begin{bmatrix} [\bar{B} \;\; F_1]\mathbf{b}_i \\ 0 \end{bmatrix}$$

5:    Solve

$$\begin{bmatrix} \sigma_i M - A^T & B^T \\ B & 0 \end{bmatrix} \begin{bmatrix} \mathbf{w}_i \\ \mathbf{q} \end{bmatrix} = \begin{bmatrix} \bar{C}^T \mathbf{c}_i \\ 0 \end{bmatrix}$$

6: **end for**

7: $V = [\mathbf{v}_1, \ldots, \mathbf{v}_\ell], W = [\mathbf{w}_1, \ldots, \mathbf{w}_\ell]$

8: $\hat{M} = W^T M V, \hat{A} = W^T A V, \hat{B} = W^T [\bar{B} \;\; F_1], \hat{C} = \bar{C} V$

9: **while** not converged **do**

10:    Compute Y, Z with $Y^* \hat{A} Z = \text{diag}(\lambda_1, \ldots, \lambda_\ell)$ and $Y^* \hat{M} Z = I_\ell$, where the columns of $Z = [\mathbf{z}_1, \ldots, \mathbf{z}_\ell]$ and $Y = [\mathbf{y}_1, \ldots, \mathbf{y}_\ell]$ are, respectively, the right   and left eigenvectors of $\lambda \hat{M} - \hat{A}$.

11:    $\sigma_i = -\lambda_i, \mathbf{b}_i^T = \mathbf{y}_i^* \hat{B}$ and $\mathbf{c}_i = \hat{C} \mathbf{z}_i$ for $i = 1, \ldots, \ell$

12:    Solve

$$\begin{bmatrix} \sigma_i M - A & B^T \\ B & 0 \end{bmatrix} \begin{bmatrix} \mathbf{v}_i \\ \mathbf{z} \end{bmatrix} = \begin{bmatrix} [\bar{B} \;\; F_1]\mathbf{b}_i \\ 0 \end{bmatrix}, \quad i = 1, \ldots, \ell$$

13:    Solve

$$\begin{bmatrix} \sigma_i M - A^T & B^T \\ B & 0 \end{bmatrix} \begin{bmatrix} \mathbf{w}_i \\ \mathbf{q} \end{bmatrix} = \begin{bmatrix} \bar{C}^T \mathbf{c}_i \\ 0 \end{bmatrix}, \quad i = 1, \ldots, \ell$$

14:    $V = [\mathbf{v}_1, \ldots, \mathbf{v}_\ell], W = [\mathbf{w}_1, \ldots, \mathbf{w}_\ell]$

15:    $\hat{M} = W^T M V, \hat{A} = W^T A V, \hat{B} = W^T [\bar{B} \;\; F_1], \hat{C} = \bar{C} V$

16: **end while**

8.2 Balanced truncation

In this section, we give a short review of balanced truncation, see [3, 35], a well-known model reduction technique for linear systems. For an approximation of dynamical systems, a small approximation error, the preservation of system properties like stability and the efficient computation of the approximate system are desirable. Balanced truncation model reduction was first introduced in [57] and [55]. In [59], it was shown that using Lyapunov based balancing, stability of the system is preserved. In [26], an error bound was introduced. For an detailed overview of model reduction by balanced truncation and other types of balancing, we refer to [3]. For systems with many inputs or outputs, balanced truncation was not efficient to use. In [10], a method was introduced to deal with this problem. This approach was also extended to descriptor systems in [11]. Here, we will present this method for the Stokes-type system (8.13)–(8.15).

8.2.1 Balanced truncation for linear time-invariant systems

First, we review balanced truncation for linear time-invariant control systems. We follow [3] and refer to [3],[35] for more details.

Consider the linear control system

$$E\frac{\mathrm{d}}{\mathrm{d}t}\mathbf{x}(t) = F\mathbf{x}(t) + G\mathbf{u}(t), \qquad (8.21)$$

$$\mathbf{y}(t) = C\mathbf{x}(t) + D\mathbf{u}(t), \qquad (8.22)$$

where $E, F \in \mathbb{R}^{n \times n}, G \in \mathbb{R}^{n \times m}, C \in \mathbb{R}^{q \times n}, D \in \mathbb{R}^{q \times m}$. We assume that the pencil $\lambda E - F$ is asymptotically stable, i.e. all its eigenvalues have negative real part. Furthermore, E is assumed to be invertible and symmetric.

The aim of model reduction is to find a reduced-order model

$$\hat{E}\frac{\mathrm{d}}{\mathrm{d}t}\hat{\mathbf{x}}(t) = \hat{F}\hat{\mathbf{x}}(t) + \hat{G}\mathbf{u}(t),$$

$$\hat{\mathbf{y}}(t) = \hat{C}\hat{\mathbf{x}}(t) + \hat{D}\mathbf{u}(t),$$

where $\hat{E}, \hat{F} \in \mathbb{R}^{\ell \times \ell}, \hat{G} \in \mathbb{R}^{\ell \times m}, \hat{C} \in \mathbb{R}^{q \times \ell}$ and $\ell \ll n$. The reduced-order model should preserve the properties of the original system, like stability.

We introduce the Lyapunov equations

$$FPE + EPF^T + GG^T = 0, \qquad (8.23)$$

$$F^TQE + EQF + C^TC = 0, \qquad (8.24)$$

where F, G and C are given system matrices from (8.21)–(8.22) and $P, Q \in \mathbb{R}^{n \times n}$ are unknown matrices. Due to the stability condition on $\lambda E - F$, these equations have a unique solution, see [3]. The solutions P and Q are called the *controllability* and *observability* Gramians, respectively.

The Lyapunov equations (8.23)–(8.24) can be solved, for example, with iterative methods like the alternating direction implicit (ADI) iteration, see [53] for more details. Knowing the solutions, we can compute the Cholesky factorization of P and Q,

$$P = UU^T,$$
$$Q = RR^T,$$

because both matrices are symmetric and positive semidefinite. The next step is the computation of the singular value decomposition (SVD) [70]

$$U^T ER = ZSY^T = \begin{bmatrix} Z_1 & Z_0 \end{bmatrix} \begin{bmatrix} S_1 & \\ & S_0 \end{bmatrix} \begin{bmatrix} Y_1^T \\ Y_0^T \end{bmatrix},$$

where $Z, Y \in \mathbb{R}^{n \times n}$ are orthogonal, $S = \mathrm{diag}(\sigma_1, \ldots, \sigma_n) \in \mathbb{R}^{n \times n}, Z_1, Y_1 \in \mathbb{R}^{n \times \ell}$ and $S_1 \in \mathbb{R}^{\ell \times \ell}$. The numbers $\sigma_i, 1 \leq i \leq n$, are called *Hankel singular values* of system (8.21)–(8.22).Without loss of generality, we assume that they are ordered decreasingly.

We define the matrices V_ℓ and W_ℓ as

$$V_\ell = U Z_1 S_1^{-\frac{1}{2}} \in \mathbb{R}^{n \times \ell},$$
$$W_\ell = R Y_1 S_1^{-\frac{1}{2}} \in \mathbb{R}^{n \times \ell},$$

and the reduced-order model is given by

$$W_\ell^T E V_\ell \frac{\mathrm{d}}{\mathrm{d}t} \hat{\mathbf{x}}(t) = W_\ell^T F V_\ell \hat{\mathbf{x}}(t) + W_\ell^T G \mathbf{u}(t), \tag{8.25}$$

$$\hat{\mathbf{y}}(t) = C V_\ell \hat{\mathbf{x}}(t) + D \mathbf{u}(t). \tag{8.26}$$

The difference between the output of the original system (8.21)–(8.22) and the reduced system (8.25)–(8.26) can be bounded by

$$\|\mathbf{y} - \hat{\mathbf{y}}\|_{L^2} \leq 2(\sigma_{\ell+1} + \cdots + \sigma_n) \|\mathbf{u}\|_{L^2}.$$

Furthermore, the stability of (8.21)–(8.22) is preserved.

8.2.2 Balanced truncation for the Stokes–Brinkman equation

In this subsection, we consider an extension of balanced truncation to Stokes-type systems developed in [66, 43, 1].

We reconsider system (8.13)–(8.15). These equations can also be written as

$$\begin{bmatrix} M & 0 \\ 0 & 0 \end{bmatrix} \frac{\mathrm{d}}{\mathrm{d}t} \begin{bmatrix} \mathbf{v} \\ \mathbf{p} \end{bmatrix} = \begin{bmatrix} A & B^T \\ B & 0 \end{bmatrix} \begin{bmatrix} \mathbf{v} \\ \mathbf{p} \end{bmatrix} + \begin{bmatrix} F_0 & F_1 \\ L & 0 \end{bmatrix} \begin{bmatrix} \mathbf{u} \\ \frac{\mathrm{d}}{\mathrm{d}t} \mathbf{u} \end{bmatrix}. \tag{8.27}$$

Due to the singularity of $\begin{bmatrix} M & 0 \\ 0 & 0 \end{bmatrix}$ the method of balanced truncation can not be applied directly to descriptor systems. It has been shown in Section 7.2, that the Stokes–Brinkman equation (8.13)–(8.15) can be written as

$$\Theta_r^T M \Theta_r \frac{\mathrm{d}}{\mathrm{d}t} \bar{\mathbf{v}}_1(t) = \Theta_r^T A \Theta_r \bar{\mathbf{v}}_1(t) + \Theta_r^T \bar{B} \mathbf{u}(t) + \Theta_r^T F_1 \frac{\mathrm{d}}{\mathrm{d}t} \mathbf{u}(t), \tag{8.28}$$

$$\mathbf{z}(t) = \bar{C} \Theta_r \bar{\mathbf{v}}_1(t) + \bar{D} \mathbf{u}(t), \tag{8.29}$$

$$\Theta_r^T M \Theta_r \bar{\mathbf{v}}_1(0) = \Theta_r^T \mathbf{v}_0, \tag{8.30}$$

where the columns of Θ_r form a basis of $\ker(B)$ and the matrices $\bar{B}, \bar{C}$ and $\bar{D}$ are given as

$$\bar{B} = F_0 - AM^{-1}B^T(BM^{-1}B^T)^{-1}L,$$
$$\bar{C} = C_1,$$
$$\bar{D} = C_1 M^{-1} B^T (BM^{-1}B^T)^{-1} L.$$

Since the matrix $\Theta_r^T M \Theta_r$ is nonsingular and the pencil $\lambda \Theta_r^T M \Theta_r - \Theta_r^T A \Theta_r$ is stable, all conditions for the application of balanced truncation for the system (8.28)–(8.30) are fulfilled, see [43]. Applying balanced truncation to system (8.28) - (8.30) we obtain the reduced-order model

$$W^T(\Theta_r^T M \Theta_r)V \frac{\mathrm{d}}{\mathrm{d}t} \hat{\mathbf{v}}_1(t) = W^T(\Theta_r^T A \Theta_r)V \hat{\mathbf{v}}_1(t) + W^T(\Theta_r^T \bar{B})\mathbf{u}(t) + W^T(\Theta_r^T F_1)\frac{\mathrm{d}}{\mathrm{d}t} \mathbf{u}(t), \tag{8.31}$$

$$\hat{\mathbf{z}}(t) = (\bar{C} \Theta_r)V \hat{\mathbf{v}}_1(t) + \bar{D} \mathbf{u}(t), \tag{8.32}$$

$$W^T(\Theta_r^T M \Theta_r)V \hat{\mathbf{v}}_1(0) = W^T \Theta_r^T \mathbf{v}_0. \tag{8.33}$$

Here the projection matrices $V = UZ_1 S_1^{-\frac{1}{2}}$ and $W = RY_1 S_1^{-\frac{1}{2}}$ are determined as in the previous subsection from the singular value decomposition

$$U^T(\Theta_r^T M \Theta_r)R = \begin{bmatrix} Z_1 & Z_0 \end{bmatrix} \begin{bmatrix} S_1 & 0 \\ 0 & S_0 \end{bmatrix} \begin{bmatrix} Y_1^T \\ Y_0^T \end{bmatrix},$$

where U and L are the Cholesky factors of the Gramians $P = UU^T$ and $Q = RR^T$. These Gramians solve the Lyapunov equations

$$(\Theta_r^T A \Theta_r)P(\Theta_r M \Theta_r) + (\Theta_r^T M \Theta_r)P(\Theta_r^T A^T \Theta_r) + (\Theta_r^T [\bar{B} \quad F_1])([\bar{B} \quad F_1]^T \Theta_r) = 0, \tag{8.34}$$

$$(\Theta_r^T A^T \Theta_r)Q(\Theta_r^T M \Theta_r) + (\Theta_r^T M \Theta_r)Q(\Theta_r^T A \Theta_r) + (\Theta_r^T \bar{C}^T)(\bar{C} \Theta_r) = 0. \tag{8.35}$$

We show now that the reduced-order system (8.31)–(8.33) can be determined without using the auxiliary system (8.28)–(8.30). For this purpose, we consider the Lyapunov equations (8.34) and (8.35).

Multiplying equations (8.34) and (8.35) from the left with Θ_l and from the right with Θ_l^T and taking into account that $\Theta_r^T \Theta_l = I$ and $\Theta_r \Theta_l^T = \Pi^T$, we get the projected Lyapunov equations

$$\Pi A \Pi^T \tilde{P} \Pi M \Pi^T + \Pi M \Pi^T \tilde{P} \Pi A^T \Pi^T + \Pi [\bar{B} \ \ F_1][\bar{B} \ \ F_1]^T \Pi^T = 0, \qquad (8.36)$$

$$\Pi A^T \Pi^T \tilde{Q} \Pi M \Pi^T + \Pi M \Pi^T \tilde{Q} \Pi A \Pi^T + \Pi \bar{C}^T \bar{C} \Pi^T = 0, \qquad (8.37)$$

with unknown matrices $\tilde{P} = \Theta_r P \Theta_r^T = \Pi^T \tilde{P} \Pi$ and $\tilde{Q} = \Theta_r Q \Theta_r^T = \Pi^T \tilde{Q} \Pi$.

Let $\tilde{P} = \tilde{U}\tilde{U}^T$ and $\tilde{Q} = \tilde{R}\tilde{R}^T$ be the Cholesky factorizations with the full-rank factors $\tilde{U}$ and $\tilde{R}$ which fulfill

$$\Pi^T \tilde{U} = \tilde{U} \text{ and } \Pi^T \tilde{R} = \tilde{R}. \qquad (8.38)$$

These relations immediately follow from

$$\Pi^T \tilde{U}\tilde{U}^T = \Pi^T \tilde{P} = \Pi^T \tilde{P} \Pi = \tilde{P} = \tilde{U}\tilde{U}^T,$$

$$\Pi^T \tilde{R}\tilde{R}^T = \Pi^T \tilde{Q} = \Pi^T \tilde{Q} \Pi = \tilde{Q} = \tilde{R}\tilde{R}^T,$$

and the fact that $\tilde{U}$ and $\tilde{R}$ are of full rank. Furthermore we obtain from

$$P = \Theta_l^T \tilde{P} \Theta_l = (\Theta_l^T \tilde{U})(\Theta_l^T \tilde{U})^T,$$

$$Q = \Theta_l^T \tilde{Q} \Theta_l = (\Theta_l^T \tilde{R})(\Theta_l^T \tilde{R})^T,$$

that the Cholesky factors U and R of $P = UU^T$ and $Q = RR^T$ are given by $U = \Theta_l^T \tilde{U}$ and $R = \Theta_l^T \tilde{R}$. Substituting these matrices in $U^T(\Theta_r^T M \Theta_r)R$ and using (8.38) we have $U^T(\Theta_r^T M \Theta_r)R = \tilde{U}^T M \tilde{R}$. Thus, the projection matrices can be determined from the singular value decomposition

$$\tilde{U}^T M \tilde{R} = \begin{bmatrix} \tilde{Z}_1 & \tilde{Z}_0 \end{bmatrix} \begin{bmatrix} \tilde{S}_1 & 0 \\ 0 & \tilde{S}_0 \end{bmatrix} \begin{bmatrix} \tilde{Y}_1^T \\ \tilde{Y}_0^T \end{bmatrix}$$

as $\tilde{V} = \tilde{U}\tilde{Z}_1\tilde{S}_1^{-1/2} \in \mathbb{R}^{n_v \times \ell}$ and $\tilde{W} = \tilde{R}\tilde{Y}_1\tilde{S}_1^{-1/2} \in \mathbb{R}^{n_v \times \ell}$ without computing the basis matrices Θ_l, Θ_r and the projector Π. Since $\Pi^T \tilde{V} = \tilde{V}$ and $\Pi^T \tilde{W} = \tilde{W}$, we obtain the reduced-order model

$$\tilde{W}^T M \tilde{V} \frac{\mathrm{d}}{\mathrm{d}t}\hat{\mathbf{v}}(t) = \tilde{W}^T A \tilde{V} \hat{\mathbf{v}}(t) + \tilde{W}^T \bar{B} \mathbf{u}(t) + \tilde{W}^T F_1 \frac{\mathrm{d}}{\mathrm{d}t}\mathbf{u}(t), \qquad (8.39)$$

$$\hat{\mathbf{z}}(t) = \bar{C}\tilde{V}\hat{\mathbf{v}}(t) + \bar{D}\mathbf{u}(t), \qquad (8.40)$$

$$\tilde{W}^T M \tilde{V} \hat{\mathbf{v}}(0) = \tilde{W}^T \mathbf{v}_0, \qquad (8.41)$$

which has the same output as (8.31)–(8.33).

The error estimate for balanced truncation can also be obtained for the Stokes–Brinkman system (8.13)–(8.15) with $\mathbf{v}_0 = 0$. The error in the output $\hat{\mathbf{z}}$ of the reduced system (8.39)–(8.41) can be estimated as

$$\|\mathbf{z} - \hat{\mathbf{z}}\|_{L^2} \le 2(\sigma_{\ell+1} + \cdots + \sigma_{n_v - n_p})\|\mathbf{u}\|_{L^2}.$$

In practice, the Gramians $\tilde{P}$ and $\tilde{Q}$ are approximated by low-rank matrices in factored form. These low-rank factors can be determined by the low-rank alternating direction implicit (LR-ADI) method. We use the variant from [9].

We initialize $W^{(0)} = \Pi[\bar{B} \ F_1], Z^{(0)} = [\ ]$. Then the ADI iteration for equation (8.36) is given by

$$V^{(k)} = (\Pi A \Pi^T + \tau_k \Pi M \Pi^T)^I W^{(k-1)}$$

and if $\mathrm{Im}(\tau_k) = 0$

$$Z^{(k)} = [Z^{(k-1)} \ \sqrt{-2\mathrm{Re}(\tau_k)} V^{(k)}],$$
$$W^{(k)} = W^{(k-1)} - 2\mathrm{Re}(\tau_k)\Pi M \Pi^T V^{(k)}.$$

Otherwise, if $\mathrm{Im}(\tau_k) \neq 0$, we have

$$Z^{(k)} = [Z^{(k-1)} \ \gamma^{(k)}(\mathrm{Re}(V^{(k)}) + \beta^{(k)}\mathrm{Im}(V^{(k)})) \ \gamma^{(k)}\sqrt{(\beta^{(k)})^2 + 1}\mathrm{Im}(V^{(k)})],$$
$$W^{(k)} = W^{(k-1)} - 4\mathrm{Re}(\tau_k)\Pi M \Pi^T(V^{\overline{(k)}} + 2\beta^{(k)}\mathrm{Im}(V^{(k)})),$$

where $\gamma^{(k)} = 2\sqrt{-\mathrm{Re}(\tau_k)}$ and $\beta^{(k)} = \frac{\mathrm{Re}(\tau_k)}{\mathrm{Im}(\tau_k)}$. The iteration is stopped if the norm of the residual $R^{(k)} = (W^{(k)})^T W^{(k)}$ is sufficiently small. Here, $(\Pi M \Pi^T + \tau_k \Pi A \Pi^T)^I$ denotes the reflexive inverse of $\Pi M \Pi^T + \tau_k \Pi A \Pi^T$ with respect to the projectors Π and Π^T. If the ADI shift parameters τ_k lie in the open left half-plane, then $Z^{(k)}(Z^{(k)})^T$ converges towards the solution of the projected Lyapunov equation (8.36) and $Z^{(k)}$ is said to be a low-rank Cholesky factor of $\tilde{P}$.

Similarly to the IRKA approach in Section 8.1, the reflexive inverses $(\Pi M \Pi^T + \tau_k \Pi A \Pi^T)^I$ do not have to be computed excplicitly. Instead, one solves linear systems of the form

$$\begin{pmatrix} M + \tau_k A & B^T \\ B & 0 \end{pmatrix} \begin{pmatrix} \mathbf{z}_1 \\ \mathbf{z}_2 \end{pmatrix} = \begin{pmatrix} \mathbf{b} \\ 0 \end{pmatrix},$$

in order to determine $\mathbf{z}_1 = (\Pi M \Pi^T + \tau_k \Pi A \Pi^T)^- \mathbf{b}$ for some vector $\mathbf{b}$.

Krylov subspace methods [46] can also be extended to the projected Lyapunov equation (8.36). For matrices F and G let

$$\mathcal{K}_k(F, G) = \mathrm{blockspan}[G, FG, \ldots, F^{k-1}G],$$

denote a block Krylov subspace. Then, similar to [67], we can compute a low-rank approximate solution $\tilde{P} \approx V_k P_k V_k^T$, where V_k has orthonormal columns that span the extended block Krylov subspace

$$\mathcal{K}_k\left((\Pi A \Pi^T)^I \Pi M \Pi^T, (\Pi A \Pi)^I \Pi[\bar{B} \ F_1]\right) + \mathcal{K}_k\left((\Pi M \Pi^T)^I \Pi A \Pi^T, (\Pi M \Pi)^I \Pi[\bar{B} \ F_1]\right),$$

and P_k solves the reduced Lyapunov equation $A_k P_k + P_k A_k^T = -B_k B_k^T$ with

$$A_k = V_k^T (\Pi M \Pi)^I \Pi A \Pi^T V_k \text{ and } B_k = V_k^T (\Pi M \Pi)^I \Pi[\bar{B} \ F_1].$$

8.2.3 Balanced truncation for systems with many inputs or outputs

The ADI method and Krylov subspace methods are efficient only for Lyapunov equations with a low-rank right-hand side. For the Stokes–Brinkman system (5.15)–(5.16) the controllability Lyapunov equation (8.36) satisfies this condition, while the observability Lyapunov equation (8.37) has the right-hand side of large rank $n_{v,1}$, since the number of outputs is large. This makes it difficult to compute the observability Gramian in a reasonable time. Balanced truncation model reduction for systems with many inputs or outputs has been proposed first in [10] for standard systems and then extended in [11] to DAEs. We will give a brief review of this method and discuss later its application to the Stokes–Brinkman equation (5.15)–(5.16).

We start with system (8.21)–(8.22) with a nonsingular, symmetric matrix E. We assume without a loss of generality that the number of inputs p is small and the number of outputs q is large. In this case the controllability Lyapunov equation (8.36) can be solved using, for example, the ADI method. As a result, we obtain the low-rank approximation $\tilde{P} = \tilde{U}\tilde{U}^T$ with a low-rank matrix $\tilde{U} \in \mathbb{R}^{n_v \times r_c}$.

For the solution of the observability Lyapunov equation (8.37), we consider the integral representation

$$Q = \frac{1}{2\pi} \int_{-\infty}^{\infty} (-i\omega E - F^T)^{-1} C^T C (i\omega E - F)^{-1} \, d\omega,$$

introduced in [65], where i denotes the complex number with $i^2 = -1$. In this representation, the direct compuation of the inverse matrices should be avoided, of course.

For the application of balanced truncation, we do not need the Cholesky factors U and R of $P = UU^T$ and $Q = RR^T$ directly, but the product $U^T E R$ whose singular value decomposition $U^T E R = Z S Y^T$ allows to determine the projection matrices. Using this singular value decomposition, we observe that

$$\begin{aligned} U^T E Q E U &= U^T E R R^T E U = (U^T E R)(U^T E R)^T \\ &= (Z S Y^T)(Z S Y^T)^T = Z S Y^T Y S^T Z^T \\ &= Z S S Z^T = Z S^2 Z^T. \end{aligned}$$

That means that the eigenvalue decomposition of $U^T E Q E U$ leads directly to the squared singular values of $U^T E R$, and the left singular vectors collected in the orthogonal matrix Z.

When we consider the matrix product $Q E U$ and replace Q by the integral identity we obtain

$$QEU = \frac{1}{2\pi} \int_{-\infty}^{\infty} (-i\omega E - F^T)^{-1} C^T C (i\omega E - F)^{-1} E U \, d\omega. \tag{8.42}$$

This integral can be approximated by a quadrature rule

$$QEU \approx \frac{1}{2\pi} \sum_{k=0}^{N} \alpha_k F(\omega_k),$$

where α_k denote the weights, ω_k the quadrature points

$$F(\omega) = U^T(-i\omega E - F^T)^{-1}C^T C(i\omega E - F)^{-1}EU.$$

Once we determine QEU, the projection matrices can be computed in the following way. The right projection matrix $V = UZ_1 S_1^{-\frac{1}{2}}$ can be computed from the eigenvalue decomposition

$$U^T EQEU = \begin{bmatrix} Z_1 & Z_0 \end{bmatrix} \begin{bmatrix} S_1^2 & \\ & S_0^2 \end{bmatrix} \begin{bmatrix} Z_1^T \\ Z_0^T \end{bmatrix}.$$

For the left projection matrix $W = RY_1 S_1^{-\frac{1}{2}}$, the matrix Y_1 with the left singular vectors of $U^T ER$ and also the Choleksy factor R are missing. But we observe that

$$QEU = R(R^T EU) = R(YS^T Z^T) = RYSZ^T = R\begin{bmatrix} Y_1 & Y_0 \end{bmatrix} \begin{bmatrix} S_1 & \\ & S_0 \end{bmatrix} \begin{bmatrix} Z_1^T \\ Z_0^T \end{bmatrix}$$

and, hence,

$$W = RY_1 S_1^{-\frac{1}{2}} = QEU Z_1 S_1^{-\frac{3}{2}}.$$

Thus, the left projection matrix W can be evaluated very easy using the integral expression for QEU and the matrices Z_1 and S_1.

We now extend this approach to the DAE system (8.13)–(8.15). First of all note that the solution of the projected Lyapunov equation (8.37) is represented as

$$\tilde{Q} = \frac{1}{2\pi}\int_{-\infty}^{\infty}(-i\omega\Pi M\Pi^T - \Pi A^T\Pi^T)^I \Pi C^T C\Pi^T(i\omega\Pi M\Pi^T - \Pi A\Pi^T)^I \, d\omega. \tag{8.43}$$

This expression can be obtained similarly to the standard state space case. Then the nonzero Hankel singular values of (8.13)–(8.15) are computed as

$$\sigma_j = \sqrt{\lambda_j(\tilde{P}\Pi M^T\Pi^T\tilde{Q}\Pi M\Pi^T)} = \sqrt{\lambda_j(\tilde{U}\tilde{U}^T M^T\tilde{Q}M\Pi^T)}$$
$$= \sqrt{\lambda_j(\tilde{U}^T M^T\tilde{Q}M\tilde{U})},$$

where $\lambda(\cdot)$ denotes the j-th eigenvalue of the corresponding matrix. Here, we used the relations $\tilde{P}\Pi = \tilde{P}, \Pi^T\tilde{U} = \tilde{U}$ and $\Pi^T\tilde{Q}\Pi = \tilde{Q}$. In this case, the projection matrices W and V can be determined using the dominant subspaces of the symmetric, positive semidefinite matrix $\tilde{U}^T M^T\tilde{Q}M\tilde{U} \in \mathbb{R}^{r_c \times r_c}$ as proposed in [10]. Let

$$\tilde{U}^T M^T\tilde{Q}M\tilde{U} = [Z_1, Z_0]\begin{bmatrix} S_1^2 & \\ & S_0^2 \end{bmatrix}[Z_1, Z_0]^T, \tag{8.44}$$

be the eigenvalue decomposition of the matrix $\tilde{U}^T M^T\tilde{Q}M\tilde{U}$, where $[Z_1, Z_0]$ is orthogonal, $S_1 = \mathrm{diag}(s_1, \ldots, s_r)$ and $S_0 = \mathrm{diag}(s_{r+1}, \ldots, s_{r_c})$ with

$$s_1 \geq \ldots \geq s_r > s_{r+1} \geq \ldots \geq s_{r_c}.$$

Then the projection matrices are determined as $V = \tilde{U} Z_1 S_1^{-\frac{1}{2}}$ and $W = \tilde{Q} M \tilde{U} Z_1 S_1^{-\frac{3}{2}}$.

Now we briefly discuss how to compute the matrix product $\tilde{Q} M \tilde{U}$ required in (8.44). Using the integral representaion (8.43) for $\tilde{Q}$ and the relations

$$C \Pi^T (i\omega \Pi M \Pi^T - \Pi A \Pi^T)^I = C(i\omega \Pi M \Pi^T - \Pi A \Pi^T)^I$$

we have

$$\tilde{Q} M \tilde{U} = \frac{1}{2\pi} \int_{-\infty}^{\infty} (-i\omega \Pi M^T \Pi^T - \Pi A^T \Pi^T)^I C^T C (i\omega \Pi M \Pi^T - \Pi A \Pi^T)^I M \tilde{U} \, d\omega. \quad (8.45)$$

This integral can be approximated by a suitable quadrature rule

$$\tilde{Q} M \tilde{U} \approx \sum_{j=1}^{\ell} \alpha_j F(i\omega_j),$$

with the weights α_j, quadrature points ω_j and

$$F_j = F(i\omega_j) = (-i\omega_j \Pi M^T \Pi^T - \Pi A^T \Pi^T)^I C^T C (i\omega_j \Pi M \Pi^T - \Pi A \Pi^T)^I M \tilde{U}. \quad (8.46)$$

This matrix-valued function can be efficiently evaluated solving the sparse linear systems

$$\begin{bmatrix} i\omega_j M - A & B^T \\ B & 0 \end{bmatrix} \begin{bmatrix} R_1 \\ R_2 \end{bmatrix} = \begin{bmatrix} M\tilde{U} \\ 0 \end{bmatrix} \quad \text{and} \quad \begin{bmatrix} -i\omega_j M^T - A^T & B^T \\ B & 0 \end{bmatrix} \begin{bmatrix} F_j \\ 0 \end{bmatrix} = \begin{bmatrix} C^T C R_1 \\ 0 \end{bmatrix}.$$

Taking into account that

$$\tilde{Q} M \tilde{U} = \frac{1}{2\pi} \int_0^\infty (F(i\omega) + \overline{F(i\omega)}) \, d\omega = \frac{1}{\pi} \int_0^\infty \operatorname{Re} F(i\omega) \, d\omega,$$

the computation of the improper integral can slightly be simplified.

Remark 8.2. Note that the symmetric matrix $\tilde{U}^T M^T \tilde{Q} M \tilde{U}$ in (8.44) can be approximated by the quadrature rule

$$\tilde{U}^T M^T \tilde{Q} M \tilde{U} \approx \sum_{j=1}^{\ell} \alpha_j ((i\omega_j \Pi M \Pi^T - \Pi A \Pi^T)^I M \tilde{U})^* (i\omega_j \Pi M \Pi^T - \Pi A \Pi^T)^I M \tilde{U},$$

where only the products $(i\omega_j \Pi M \Pi^T - \Pi A \Pi^T)^I M \tilde{U}$ need to be computed. However, we still need $\tilde{Q} M \tilde{U}$. Therefore, we prefer to work with (8.43) from the beginning.

8.3 POD for model reduction of nonlinear equations

In this section, we describe the Proper Orthogonal Decomposition (POD) method [69], [70], [13] for model reduction of nonlinear systems. This method was first conceived for continuous second-order processes, see [13] for an overview and is also known as Karhunen-Loève decomposition [49],[51],[52],[58],[60]. Restricted to a finite-dimensional case, POD is also well known and used as Principal Component Analysis (PCA) in statistics [47], [64]. In [69], [70] POD has been applied to systems of ordinary differential equations in order to construct a lower dimensional problem which approximates the orginial equations. For a more detailed overview of the history of POD, we refer to [50].

We will follow [69] for the notation.

8.3.1 Problem formulation and snapshots

We consider a nonlinear system of ordinary differential equations of the form

$$
\begin{aligned}
M\dot{\mathbf{y}}(t) &= A\mathbf{y}(t) + \mathbf{f}(\mathbf{y}(t)), \\
M\mathbf{y}(0) &= \mathbf{y}_0,
\end{aligned}
\tag{8.47}
$$

where $M, A \in \mathbb{R}^{n \times n}$, $\mathbf{y}_0 \in \mathbb{R}^n$ and the nonlinearity $\mathbf{f}: \mathbb{R}^n \to \mathbb{R}^n$ are given and the solution $\mathbf{y}: [0, T] \to \mathbb{R}^n$ should be determined. If the original dimension n of the problem is very large, then the computation of the solution is very expensive. Therefore an approximation of (8.47) by a model of lower dimension is required in order to reduce the computational complexity of solving the system.

The idea is to find a low-dimensional subspace which contains approximately the solution trajectory of (8.47). For this purpose, we use POD based on a Galerkin projection, where the basis vectors $\mathbf{w}_1, \ldots, \mathbf{w}_\ell \in \mathbb{R}^n$ solve the minimization problem:

Minimize

$$
\sum_{i=1}^{m} \left\| \mathbf{y}(t_i) - \sum_{j=1}^{\ell} (\mathbf{y}(t_i), \mathbf{w}_j)_M \mathbf{w}_j \right\|_2^2
\tag{8.48}
$$

subject to $(\mathbf{w}_j, \mathbf{w}_k)_2 = \delta_{jk}, 1 \leq j, k \leq \ell$.

In other words, we search for an orthogonal basis of an ℓ-dimensional subspace of $\mathbb{R}^n$ which contains approximately the solution snapshots $\mathbf{y}(t_1), \ldots, \mathbf{y}(t_m)$. These snapshots have to be in a certain way characteristic for the whole solution which means that a proper choice of them is very important. There are several suggestions how the snapshots can be chosen, for example using the time steps of a time discretization method for solving the system. It is also possible to add information of the derivative to these snapshots and to use different weights for them [69].

In order to use POD we already have to know the solution at the times $t_1, \ldots, t_m$. If we have to solve a differential equation only once, this technique is not worth the effort. However in optimization problems governed by partial or ordinary differential equations, we are supposed to solve these equations with varying parameters in each step of the optimization process. In this case, solving the reduced systems instead could reduce the computation time significantly.

8.3.2 Optimal solution by SVD

In this subsection we discuss how the optimization problem (8.48) can be solved. Let, therefore, $Y = [\mathbf{y}_1, \ldots, \mathbf{y}_m] \in \mathbb{R}^{n \times m}$ denote the matrix of snaphots with $\mathbf{y}_i = \mathbf{y}(t_i)$, for $i = 1, \ldots, m$, where m is assumed to be smaller than n. We consider the SVD

$$Y = W \Sigma V^T,$$

where $W \in \mathbb{R}^{n \times n}$ and $V \in \mathbb{R}^{m \times m}$ are orthogonal matrices. Furthermore, $\Sigma = \mathrm{diag}(\sigma_1, \ldots, \sigma_m) \in \mathbb{R}^{n \times m}$ is a diagonal matrix with $\sigma_1 \geq \ldots \geq \sigma_m \geq 0$. It was proved in [70] that the first ℓ columns of W solve the optimization problem (8.48). That means, in practice, we have to compute the singular value decomposition of the matrix of snapshots in order to determine a POD basis $\mathbf{w}_1, \ldots, \mathbf{w}_\ell$ solving (8.48). Note that (8.48) has no unique solution, as the order or the signs of the POD basis vectors can be changed. Any other basis of $\mathrm{span}\{\mathbf{w}_1, \ldots, \mathbf{w}_\ell\}$, will also solve the minimization problem (8.48).

In [70], an alternative way for the computation of a POD basis was also presented which will be used later on. Therefore, we give a short review of this variant. Consider the symmetric positive definite matrix $\mathcal{K} \in \mathbb{R}^{m \times m}$, where

$$\mathcal{K}_{i,j} = (\mathbf{y}_j, \mathbf{y}_i)_2.$$

There exists an eigenvalue decomposition

$$\mathcal{K} = V D V^T,$$

where $V = [\mathbf{v}_1, \ldots, \mathbf{v}_m] \in \mathbb{R}^{m \times m}$ and $D = \mathrm{diag}(\lambda_1, \ldots, \lambda_m) \in \mathbb{R}^{m \times m}$ with the eigenvalues $\lambda_i, i = 1, \ldots, m$. Then, in [70] it was shown that the POD basis can be written as

$$\mathbf{w}_i = \frac{1}{\sqrt{\lambda_i}} Y \mathbf{v}_i, i = 1, \ldots, \ell,$$

with $\lambda_i \neq 0$.

8.3.3 Model reduction by projection

Once a POD basis $\mathbf{w}_1, \ldots, \mathbf{w}_\ell$ is determined, one can choose as ansatz

$$\mathbf{y}(t) \approx \sum_{j=1}^{\ell} \alpha_j(t) \, \mathbf{w}_j = W_\ell \hat{\mathbf{y}}(t),$$

where $W_\ell := [\mathbf{w}_1, \ldots, \mathbf{w}_\ell] \in \mathbb{R}^{n \times \ell}$ und $\hat{\mathbf{y}}(t) = [\alpha_1(t), \ldots, \alpha_\ell(t)]^T \in \mathbb{R}^\ell$. Due to the assumption, that the solution lives approximately in the span of the POD basis, this ansatz makes sense. Inserting this approximation in (8.47) and multiplying the resulting equation from the left by W_ℓ^T we get a reduced-order model

$$\begin{aligned} \hat{M} \dot{\hat{\mathbf{y}}}(t) &= \hat{A} \hat{\mathbf{y}}(t) + \hat{\mathbf{f}}(\hat{\mathbf{y}}(t)), \\ \hat{M} \hat{\mathbf{y}}(0) &= \hat{\mathbf{y}}_0, \end{aligned} \tag{8.49}$$

where

$$\hat{M} = W_\ell^T M W_\ell \in \mathbb{R}^{\ell \times \ell},$$
$$\hat{A} = W_\ell^T A W_\ell \in \mathbb{R}^{\ell \times \ell},$$
$$\hat{\mathbf{f}}(\hat{\mathbf{y}}(t)) = W_\ell^T \mathbf{f}(W_\ell \hat{\mathbf{y}}(t)) \in \mathbb{R}^\ell, \text{ and}$$
$$\hat{\mathbf{y}}_0 = W_\ell^T \mathbf{y}_0 \in \mathbb{R}^\ell.$$

Note that, even if the matrices M, A of the original system are sparse, the matrices $\hat{M}$, $\hat{A}$ of the reduced-order model are, in general, not.

In order to compute $\hat{\mathbf{f}}(\hat{\mathbf{y}}(t))$, it is required first to compute $W_\ell \hat{\mathbf{y}}(t)$ (which costs $O(n\ell)$ operations), then the evaluation of the large vector $\mathbf{f}(W_\ell \hat{\mathbf{y}}(t))$ has to be done and, finally, the multiplication with W_ℓ^T (which costs also $O(n\ell)$ operations). Thus, due to the nonlinearity, the computational complexity of the reduced system still depends on the dimension n of the original problem.

8.3.4 Discrete empirical interpolation method

The discrete empirical interpolation method (DEIM) for fast evaluation of the nonlinearity was developed in [17]. Here, we give a short review of this method.

We assume again that the nonlinearity $\mathbf{f}(\mathbf{y}(t))$ lives approximately in a low-dimensional subspace $\mathcal{U} \subset \mathbb{R}^n$ and search for a projection onto this subspace which offers an efficient evaluation of the projected nonlinearity. With snapshots $\mathbf{f}(\mathbf{y}(t_1)), \ldots, \mathbf{f}(\mathbf{y}(t_m))$ at the times $t_1, \ldots, t_m$ we can use again POD in order to find an orthonormal basis $\mathbf{u}_1, \ldots, \mathbf{u}_k$ of the subspace $\mathcal{U} \subset R^n$. Let

$$F = [\mathbf{f}(\mathbf{y}(t_1)), \ldots, \mathbf{f}(\mathbf{y}(t_m))] = U_F \Sigma_F V_F^T \tag{8.50}$$

be the singular value decomposition of F, $U_F = [\mathbf{u}_1, \ldots, \mathbf{u}_m]$ and V_F have orthogonal columns and Σ_F is diagonal whose diagonal elements are ordered decreasingly. Introducing the POD basis matrix $U_k = [\mathbf{u}_1, \ldots, \mathbf{u}_k] \in \mathbb{R}^{n \times k}$ we use the ansatz

$$U_k \mathbf{c} \approx \mathbf{f}(W_\ell \hat{\mathbf{y}}(t)),$$

where the vector $\mathbf{c} \in \mathbb{R}^k$ contains the coefficients for the representation of the vector-valued function $\mathbf{f}(W_\ell \hat{\mathbf{y}}(t))$ as a linear combination of the basis vectors $\mathbf{u}_1, \ldots, \mathbf{u}_k$. Since we can not guarantee to fulfill n conditions while having k unknowns, we select k special equations which should hold true, i.e.

$$P^T U_k \mathbf{c} = P^T \mathbf{f}(W_\ell \hat{\mathbf{y}}(t)), \tag{8.51}$$

with a selector matrix $P = [\mathbf{e}_{\varphi_1}, \ldots, \mathbf{e}_{\varphi_k}] \in \mathbb{R}^{n \times k}$. Here, $\varphi_1, \ldots, \varphi_k$ are the indices of the chosen equations and $\mathbf{e}_j$ denotes the jth column of the identity matrix.

If $P^T U_k$ is invertible, the coefficient vector $\mathbf{c}$ can be determined uniquely as

$$\mathbf{c} = (P^T U_k)^{-1} P^T \mathbf{f}(W_\ell \hat{\mathbf{y}}(t)).$$

We finally end up with

$$\mathbf{f}(W_\ell \hat{\mathbf{y}}(t)) \approx U_k \mathbf{c} = U_k (P^T U_k)^{-1} P^T \mathbf{f}(W_\ell \hat{\mathbf{y}}(t)).$$

Here, $\mathcal{Q} := U_k (P^T U_k)^{-1} P^T \in \mathbb{R}^{n \times n}$ is indeed a projector, since

$$\mathcal{Q}^2 = U_k ((P^T U_k)^{-1} P^T U_k)(P^T U_k)^{-1} P^T = U_k (P^T U_k)^{-1} P^T = \mathcal{Q}.$$

The DEIM indices can be computed by a greedy procedure presented in Algorithm 4. The input for the algorithm is the POD basis $\mathbf{u}_1, \ldots, \mathbf{u}_k$. In the first step, we choose the index of the largest component of the first basis vector $\mathbf{u}_1$ with respect to the absolut value as the first index φ_1. The singular value decomposition (8.50) returns the basis vectors sorted by importance, so the corresponding φ_1-th component of $\mathbf{f}(W_\ell \hat{\mathbf{y}}(t))$ is globally the most important one. In the further steps, we determine the residual $\mathbf{r}$ of the expression of the next basis vector by the previous ones and choose the next index as the index of the largest component of the residual.

In case that system (8.47) is the result of a finite difference or finite element discretization, the indices of the selected k equations have a vivid meaning. Then the components of the vector $\mathbf{f}(W_\ell \hat{\mathbf{y}}(t))$ correspond to points of the mesh and degrees of freedom, respectively. By the special construction of the projector, the nonlinearity is interpolated in the sense that the φ_1- to φ_k-coordinates of $\mathbf{f}(W_\ell \hat{\mathbf{y}}(t))$ and $\tilde{\mathbf{f}}(W_\ell \hat{\mathbf{y}}(t)) = U_k (P^T U_k)^{-1} P^T \mathbf{f}(W_\ell \hat{\mathbf{y}}(t))$ agree, since $P^T \mathbf{f}(W_\ell \hat{\mathbf{y}}(t)) = P^T \tilde{\mathbf{f}}(W_\ell \hat{\mathbf{y}}(t))$.

For the evaluation of $P^T \mathbf{f}(W_\ell \hat{\mathbf{y}}(t))$ only the k-components of $\mathbf{f}(W_\ell \hat{\mathbf{y}}(t))$ have to be determined, the computation of the other ones is not necessary. But the reduction of the computational costs itself depends on the form of the nonlinearity. Often, systems like (8.47) come from a finite difference or finite element discretization of a partial differential equation. Partial differential equations deal with local information of their solutions, like derivatives. Hence, in practice, the entries of $\mathbf{f}$ only depend on a few entries of the vector $W_\ell \hat{\mathbf{y}}$. Assuming this, we do not have to compute the whole vector $W_\ell \hat{\mathbf{y}}$ but instead just a few entries and, therefore, the computational complexity does not depend on the size of the original problem any longer.

In order to solve the linear systems in Algorithm 4, we need in each step that the corresponding matrix $P^T U$ is regular. The following theorem from [17] shows that these matrices are nonsingular and gives a bound for the DEIM error $\|\mathbf{f}(W_\ell \hat{\mathbf{y}}(t)) - \tilde{\mathbf{f}}(W_\ell \hat{\mathbf{y}}(t))\|_2$.

Theorem 8.3. *(a) In each step of Algorithm 4 the matrix $P^T U$ is nonsingular.*

(b) For the approximation

$$\tilde{\mathbf{f}}(W_\ell \hat{\mathbf{y}}(t)) := U (P^T U)^{-1} P^T \mathbf{f}(W_\ell \hat{\mathbf{y}}(t))$$

Algorithm 4 Greedy procedure for computing the DEIM indices

1: **Input:** $\mathbf{u}_1, \ldots, \mathbf{u}_k \in \mathbb{R}^n$ linearly independent
2: **Output:** $\boldsymbol{\varphi} = (\varphi_1, \ldots, \varphi_k) \in \mathbb{R}^{1 \times k}$
3: $\varphi_1 := \arg\max_{i=1,\ldots,n} |(\mathbf{u}_1)_i|$
4: $\quad U \leftarrow (\mathbf{u}_1), \; P \leftarrow (\mathbf{e}_{\varphi_1}), \; \boldsymbol{\varphi} \leftarrow (\varphi_1)$
5: **for** $l = 2, \ldots, k$ **do**
6: $\quad$ Determine $\mathbf{c}$ from $(P^T U)\mathbf{c} = P^T \mathbf{u}_l$.
7: $\quad$ $\mathbf{r} = \mathbf{u}_l - U\mathbf{c}$
8: $\quad$ $\varphi_l = \arg\max_{i=1,\ldots,n} |(\mathbf{r})_i|$
9: $\quad$ $U \leftarrow [U, \mathbf{u}_l], \; P \leftarrow [P, \mathbf{e}_{\varphi_l}], \; \boldsymbol{\varphi} \leftarrow [\boldsymbol{\varphi}, \varphi_l]$
10: **end for**

---

the DEIM error can be bounded as

$$\|\mathbf{f}(W_\ell \hat{\mathbf{y}}(t)) - \tilde{\mathbf{f}}(W_\ell \hat{\mathbf{y}}(t))\|_2 \leq \|(P^T U)^{-1}\|_2 \, \|(I - UU^T)\mathbf{f}(W_\ell \hat{\mathbf{y}}(t))\|_2,$$

where $\|(I - UU^T)\mathbf{f}(W_\ell \hat{\mathbf{y}}(t))\|_2$ *is the POD error.*

In [17] a theoretical upper bound is derived for the condition number of $P^T U$, but the bound is very conservative.

8.3.5 Matrix DEIM

The goal in this subsection is to extend DEIM to time-dependent matrices as they appear in the advection-diffusion equation (7.77)–(7.78) and almost all of the numerical costs are hidden in the assembling of these matrices. DEIM had already been applied to matrices in [17] in order to handle Jacobi matrices.

We aim to find an approximation for the matrix-valued function $M : \mathbb{R}^q \to \mathbb{R}^{n \times n}$ in the following way

$$M(\mathbf{x}) \approx \sum_{j=1}^{k} \tilde{f}_j(\mathbf{x}) U_j, \tag{8.52}$$

where $U_j \in \mathbb{R}^{n \times n}$ are constant matrices and $\tilde{f}_j : \mathbb{R}^q \to \mathbb{R}$ are continuous functions. Such an approximation can be determined by the DEIM method applied to the vector-valued function $\text{vec}(M(\mathbf{x}))$, obtained from the matrix $M(\mathbf{x})$ by stacking its columns one after another. Collecting the snapshots $\mathbf{m}_1 = \text{vec}(M(\mathbf{x}_1)), \ldots, \mathbf{m}_r = \text{vec}(M(\mathbf{x}_r))$ we determine the POD basis $\mathbf{u}_1, \ldots, \mathbf{u}_k$ from the singular value decomposition of $[\mathbf{m}_1, \ldots, \mathbf{m}_r]$. Using this basis we can determine a selector matrix $\mathcal{P}$ by Algorithm 4 and compute the approximation $\text{vec}(M(\mathbf{x})) \approx U(\mathcal{P}^T U)^{-1}\mathcal{P}^T \text{vec}(M(\mathbf{x}))$ with $U = [\mathbf{u}_1, \ldots, \mathbf{u}_k]$. Then the approximation (8.52) is obtained by reversing the vectorization. As a result, we get (8.52) with $U_j = \text{vec}^{-1}(\mathbf{u}_j)$ and $\tilde{f}_j = ((\mathcal{P}^T U)^{-1}\mathcal{P}^T \text{vec}(M(\mathbf{x})))_j$.

Our goal is now to reformulate this approximation method in order to get rid of the vectorization. We want to avoid the need of storing sparse matrices additionally in vector form.

In order to reformulate DEIM, we make some general observations. Similar to before, we collect first the snapshots $M_1 = M(\mathbf{x}_1), \ldots, M_r = M(\mathbf{x}_r)$ for the parameters $\mathbf{x}_1, \ldots, \mathbf{x}_r$. In the next step, a POD basis $(U_j)_{j=1}^{k} \in \mathbb{R}^{n \times n}$ has to be obtained. Note that in this case the POD basis consists of matrices and not of vectors. Therefore, the alternative method of computing a POD basis reviewed in subsection 8.3.2 comes into play. We compute the eigenvalue decomposition of the matrix $\mathcal{K} = ((M_i, M_j)_F)_{i,j=1}^{r}$ where $(\cdot, \cdot)_F$ denotes the Frobenius inner product of matrices.

Note that $\mathcal{K} \in \mathbb{R}^{r \times r}$ is in general a small matrix, because r is the number of snapshots which is in general much smaller than the dimension of the problem. Hence, computing the eigenvalue decomposition can be easily obtained.

Next, we consider the expression $\mathcal{P}^T \mathrm{vec}(M(\mathbf{x})) =: \mathcal{P}^T \mathbf{z}$ which shows up in the approximation. Let $r_1, \ldots, r_k$ denote the indices which are chosen by the DEIM selection algorithm. Then it holds

$$
\mathcal{P}^T \mathrm{vec}(M(\mathbf{x})) = \begin{pmatrix} \mathbf{z}_{r_1} \\ \vdots \\ \mathbf{z}_{r_k} \end{pmatrix} = \begin{pmatrix} M(\mathbf{x})_{i_1, j_1} \\ \vdots \\ M(\mathbf{x})_{i_k, j_k} \end{pmatrix}, \tag{8.53}
$$

where $r_l = n(j_l - 1) + i_l$. As a result, this expression can be computed without making use of the vectorization, because instead of selecting indices of a vector, obviously row and column indices of a matrix can be selected in the matrix DEIM implementation without any additional effort.

Furthermore we take the matrix $\mathcal{P}^T U \in \mathbb{R}^{k \times k}$ into account, which also appears in the approximation. This matrix has the form

$$
\mathcal{P}^T U = \begin{pmatrix} (\mathbf{u}_1)_{r_1} & \cdots & (\mathbf{u}_k)_{r_1} \\ \vdots & \ddots & \ldots \\ (\mathbf{u}_1)_{r_k} & \cdots & (\mathbf{u}_k)_{r_k} \end{pmatrix}.
$$

Using the row and column indices (i_l, j_l) as in (8.53) it can be rewritten as

$$
\mathcal{P}^T U = \begin{pmatrix} (U_1)_{i_1, j_1} & \cdots & (U_k)_{i_1, j_1} \\ \vdots & \ddots & \ldots \\ (U_1)_{i_k, j_k} & \cdots & (U_k)_{i_k, j_k} \end{pmatrix}.
$$

Hence, also this matrix can be obtained without the utilization of the vec-operator.

Introducing $\mathbf{c} := (\mathcal{P}^T U)^{-1} \mathcal{P}^T \mathrm{vec}(M(\mathbf{x}))$, the vector $U\mathbf{c}$ has to be computed and re-vectorized in order to obtain the approximation matrix (8.52). Due to the linearity of vec^{-1}, it holds

$$
\mathrm{vec}^{-1}(U\mathbf{c}) = \mathrm{vec}^{-1}(\sum_{i=1}^{k} \mathbf{u}_i \mathbf{c}_i) = \sum_{i=1}^{k} \mathbf{c}_i \mathrm{vec}^{-1}(\mathbf{u}_i) = \sum_{i=1}^{k} \mathbf{c}_i U_i,
$$

which means that the vectorization operator is not needed in the computation of the approximation matrix, if the vector $\mathbf{c}$ is known.

Finally, we consider the DEIM selection algorithm. For the residuals, which are vectorized matrices, in the step l of Algorithm 4 the expression

$$U(\mathcal{P}^T U)^{-1}\mathcal{P}^T \mathbf{u}_l,$$

is required to be computed, where $U = [\mathbf{u}_1, \ldots, \mathbf{u}_{l-1}], \mathcal{P} = [\mathbf{e}_{r_1}, \ldots, \mathbf{e}_{r_{l-1}}] \in \mathbb{R}^{n^2 \times (l-1)}$. In fact, for the determination of this expression, no vectorization is needed. We notice that it holds

$$\mathbf{b} = \mathcal{P}^T \mathbf{u}_l = \begin{pmatrix} (U_l)_{i_1,j_1} \\ \vdots \\ (U_l)_{i_{l-1},j_{l-1}} \end{pmatrix}.$$

This leads to the fact that the matrix $G := \mathcal{P}^T U$ can be computed succesively, by adding one row and column in each iteration

$$\left[\begin{array}{c|c} G & \mathbf{b} \\ \hline (U_1)_{i_l,j_l} \cdots (U_{l-1})_{i_l,j_l} & (U_l)_{i_l,j_l} \end{array} \right].$$

This saves computational costs.

The greedy procedure for determining the matrix DEIM indices is summarized in Algorithm 5. Note that the DEIM error estimate holds also true for matrix DEIM, which can be easily obtained by using the equivalent formulation with the vec-operator.

Algorithm 5 Greedy procedure for computing matrix DEIM indices
---
1: **Input:** $U_1, \ldots, U_k \in \mathbb{R}^{n \times n}$
2: **Output:** $G, (i_1, j_1), \ldots, (i_k, j_k)$
3: $(i_1, j_1) = \arg\max_{i=1,\ldots,n,j=1,\ldots,n} |(U_1)_{i,j}|$
4: $G = (U_1)_{i_1,j_1}$
5: **for** $l = 2, \ldots, k$ **do**
6:     Determine $\mathbf{c}$ from $G\mathbf{c} = \mathbf{b}$ with $\mathbf{b} = \begin{bmatrix} (U_l)_{i_1,j_1} \\ \vdots \\ (U_l)_{i_{l-1},j_{l-1}} \end{bmatrix}$.
7:     $R = U_l - \sum_{i=1}^{l-1} U_i(\mathbf{c})_i$
8:     $(i_l, j_l) = \arg\max_{i=1,\ldots,n,j=1,\ldots,n} |(R)_{i,j}|$
9:     $G \leftarrow \left[\begin{array}{c|c} G & \mathbf{b} \\ \hline (U_1)_{i_l,j_l} \cdots (U_{l-1})_{i_l,j_l} & (U_l)_{i_l,j_l} \end{array} \right]$
10: **end for**

Furthermore it can be noticed that DEIM can be reformulated in the same way in a general, finite dimensional Hilbert space, if there is an explicitly given basis to work with.

8.3.6 Model reduction for the advection-diffusion equation

In this subsection, we dicuss model reduction of the advection-diffusion equation (7.77)–(7.78) using a combination of POD and DEIM. We will see that the reduced-order model obtained by using POD is still very expensive to solve. This problem arises due to the dependence of the system matrix on the velocity vector coming from the Stokes–Brinkman equation. When the velocity changes, the system matrix has to be recomputed. Since the assembling of this matrix is quite expensive, we will employ the matrix DEIM approach to reduce the computational complexity.

Consider the semidiscretized advection-diffusion equation

$$M_{ad}\dot{\mathbf{c}}(t) = A_{ad}(\mathbf{v}(t))\mathbf{c}(t),$$
$$M_{ad}\mathbf{c}(0) = \mathbf{c}_0,$$

where $\mathbf{v}(t) \in \mathbb{R}^{n_v}$ is the velocity vector from the Stokes–Brinkman equation (8.13)–(8.15), $A_{ad}(\mathbf{v}(t)) \in \mathbb{R}^{n_c \times n_c}$ and $\mathbf{c}_0 \in \mathbb{R}^{n_c}$.

For the snapshots $\mathbf{c}_1 = \mathbf{c}(t_1), \ldots, \mathbf{c}_q = \mathbf{c}(t_q)$ at the times $t_1, \ldots, t_q$ with $q \leq n_c$, we compute the singular value decomposition

$$X_c = [\mathbf{c}_1, \ldots, \mathbf{c}_q] = W_c \Sigma_c V_c^T,$$

where $W_c = [\mathbf{w}_1, \ldots, \mathbf{w}_{n_c}] \in \mathbb{R}^{n_c \times n_c}, V_c \in \mathbb{R}^{q \times q}$ have orthogonal columns and the diagonal elements of $\Sigma_c = \text{diag}(\sigma_1, \ldots, \sigma_q) \in \mathbb{R}^{n_c \times q}$ are ordered decreasingly.

Then the projection matrix $W_\ell \in \mathbb{R}^{n_c \times \ell}$ is chosen as

$$W_\ell = [\mathbf{w}_1, \ldots, \mathbf{w}_\ell].$$

Finally, we derive the POD-reduced system

$$\hat{M}_{ad}\dot{\hat{\mathbf{c}}}(t) = \hat{A}_{ad}(\mathbf{v}(t))\hat{\mathbf{c}}(t), \tag{8.54}$$
$$\hat{M}_{ad}\hat{\mathbf{c}}(0) = \hat{\mathbf{c}}_0, \tag{8.55}$$

where $\hat{M}_{ad} = W_\ell^T M_{ad} W_\ell, \hat{A}_{ad}(\mathbf{v}(t)) = W_\ell^T A_{ad}(\mathbf{v}(t))W_\ell \in \mathbb{R}^{\ell \times \ell}$ and $\hat{\mathbf{c}}(t) \in \mathbb{R}^\ell$.

In system (8.54)–(8.55), the matrix $A_{ad}(\mathbf{v})$ has to be assembled at every time step. This is usually the most expensive computational part, meaning that POD itself does not help to reduce the effort at all. Therefore, we apply matrix DEIM for the matrix A_{ad}. We have to compute snapshots $A_1 = A_{ad}(\mathbf{v}(t_1)), \ldots, A_q = A_{ad}(\mathbf{v}(t_q))$ at the times $t_1, \ldots, t_q$. From the eigenvalue decomposition of the matrix $\mathcal{K} = ((A_i, A_j)_F)_{i,j=1}^q$ we obtain a POD basis $U_1, \ldots, U_k$. The application of Algorithm 5 gives us the matrix G and thus, we obtain the approximation

$$A_{ad}(\mathbf{v}(t)) \approx \sum_{j=1}^k (G^{-1}\mathbf{a}(\mathbf{v}(t)))_j U_j,$$

where $\mathbf{a}(\mathbf{v}(t)) = \begin{bmatrix} A_{ad}(\mathbf{v}(t))_{i_1,j_1} \\ \vdots \\ A_{ad}(\mathbf{v}(t))_{i_k,j_k} \end{bmatrix}$, with the selected indices i_l, j_l from Algorithm 5. Then, the reduced-order model reads

$$\hat{M}_{ad}\dot{\hat{\mathbf{c}}}(t) = \sum_{j=1}^{k} (G^{-1}\mathbf{a}(\mathbf{v}(t)))_j W_\ell^T U_j W_\ell \hat{\mathbf{c}}(t), \tag{8.56}$$

$$\hat{M}_{ad}\hat{\mathbf{c}}(0) = \hat{\mathbf{c}}_0. \tag{8.57}$$

All in all, we have to precompute the projection matrix W_ℓ using POD as well as the POD basis matrices U_j, the matrix G from Algorithm 5 and the selection indices (i_l, j_l). The matrices $W_\ell^T U_j W_\ell \in \mathbb{R}^{\ell \times \ell}$ can also be precomputed. If we apply the numerical method to (8.56)–(8.57), we only have to compute the k entries $A_{ad}(\mathbf{v}(t))_{i_1,j_1}, \ldots, A_{ad}(\mathbf{v}(t))_{i_k,j_k}$ of $A_{ad}(\mathbf{v}(t))$ and $G^{-1}\mathbf{a}(\mathbf{v}(t))$ at each time step.

8.4 The reduced gradient method

Using model order reduction we can reformulate the gradient method. Note that the adjoint and state Stokes–Brinkman equations as well as the adjoint and state advection-diffusion equations have a similar structure. That means we can compute reduced-order models for these related equations at a stroke. Figure 8.1 shows the coupled system of the reduced equations.

The reduced system of equations reads

$$\hat{M}\frac{\mathrm{d}}{\mathrm{d}t}\hat{\mathbf{v}}(t) = \hat{A}\hat{\mathbf{v}}(t) + \hat{B}_1\mathbf{u}(t) + \hat{B}_2\frac{\mathrm{d}}{\mathrm{d}t}\mathbf{u}(t), \quad \hat{M}\hat{\mathbf{v}}(0) = \hat{\mathbf{v}}_0, \tag{8.58}$$

$$\hat{\mathbf{z}}(t) = \hat{C}\hat{\mathbf{v}}(t) + \hat{D}\mathbf{u}(t), \tag{8.59}$$

$$\hat{M}_{ad}\frac{\mathrm{d}}{\mathrm{d}t}\hat{\mathbf{c}}(t) = \hat{A}_{ad}(\hat{\mathbf{z}}(t))\hat{\mathbf{c}}(t), \tag{8.60}$$

$$\hat{M}_{ad}\hat{\mathbf{c}}(0) = \hat{\mathbf{c}}_0, \tag{8.61}$$

$$-\hat{M}_{ad}\frac{\mathrm{d}}{\mathrm{d}t}\hat{\mathbf{d}}(t) = (\hat{A}_{ad}(\hat{\mathbf{z}}(t)))^T\hat{\mathbf{d}}(t), \tag{8.62}$$

$$\hat{\mathbf{d}}(T) = \hat{\mathbf{c}}^{\text{foc}} - \hat{\mathbf{c}}(T), \tag{8.63}$$

$$\hat{\mathbf{w}} = \hat{\mathcal{F}}_{\hat{\mathbf{z}},\hat{\mathbf{c}}}^*\hat{\mathbf{d}}, \tag{8.64}$$

$$-\hat{M}\frac{\mathrm{d}}{\mathrm{d}t}\hat{\boldsymbol{\lambda}}(t) = \hat{A}^T\hat{\boldsymbol{\lambda}}(t) + \hat{C}^T\hat{\mathbf{w}}(t), \tag{8.65}$$

$$\hat{M}\hat{\boldsymbol{\lambda}}(T) = 0, \tag{8.66}$$

$$\hat{\mathbf{g}}(t) = \mathcal{P}_\Upsilon(\sigma\mathbf{u}(t) - \hat{B}_1^T\hat{\boldsymbol{\lambda}}(t) + \hat{B}_2^T\frac{\mathrm{d}}{\mathrm{d}t}\hat{\boldsymbol{\lambda}}(t) + \bar{D}^T\hat{\mathbf{w}}). \tag{8.67}$$

Algorithm 6 summarizes the reduced gradient method.

Algorithm 6 Reduced gradient method

1: **Input:** $\mathbf{u}_{\text{start}}, \epsilon_{\text{tol}}$
2: **Output:** $\mathbf{u}^*$
3: Initialization: $\mathbf{u}^{(0)} = \mathbf{u}_{\text{start}}, k = 0$
4: **while** $\|\hat{\mathbf{g}}\| > \epsilon_{\text{tol}}$ **do**
5:    Solve the reduced Stokes–Brinkman equation (8.58)–(8.59) for $\hat{\mathbf{z}}$.
6:    Solve the reduced advection-diffusion equation (8.60)–(8.61) for $\hat{\mathbf{c}}$.
7:    Solve the reduced advection-diffusion equation (8.62)–(8.63) for $\hat{\mathbf{d}}$.
8:    Solve the reduced Stokes–Brinkman equation (8.65)–(8.66) for $\hat{\boldsymbol{\lambda}}$.
9:    Compute the gradient $\hat{\mathbf{g}}$ from (8.67).
10:    Compute a step size α using Armijo line search
11:    $\mathbf{u}^{(k+1)} = \mathbf{u}^{(k)} - \alpha\hat{\mathbf{g}}$
12:    $k = k + 1$
13: **end while**

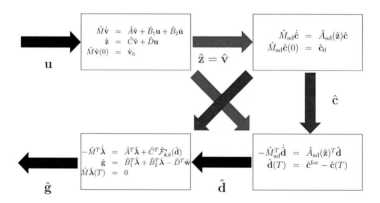

Figure 8.1: The coupled system of reduced equations.

9 Numerical experiments

In this chapter, we present our numerical results. First, we compare the results of the model reduction techniques we used and afterwards we discuss the optimal control problem.

9.1 Model reduction results

9.1.1 Stokes–Brinkman equation and its adjoint

The original dimensions were chosen as $n_v = 4440$, $n_p = 621$. First, we discuss the results using balanced truncation for many outputs. In Figure 9.1, the error of the transfer functions in the frequency domain is plotted. Furthermore, Figure 9.2 shows the relative error in the output of the Stokes–Brinkman equation. The reduced dimension was $\ell = 300$. While the fineness of the model reduction using balanced truncation is not too impressive, the application of IRKA led to much better results. Furthermore, the implementation of balanced truncation is too expensive. Figure 9.4 shows the error in the frequency response in the IRKA algorithm, while in Figure 9.3 the relative error of the output in both state and adjoint Stokes–Brinkman equation is plotted. The reduced dimension was $\ell = 60$. The following table compares both methods in terms of computational effort.

Time for solving the state equation in seconds	3.0381
Time for solving the adjoint equation in seconds	3.1473

	IRKA	Balanced truncation
Solving reduced state equation in seconds	0.0149	0.1321
Solving reduced adjoint equation in seconds	0.1211	0.0067
Time for computing the reduced order models in seconds	110.8337	1582.4899

9.1.2 The advection-diffusion equation and its adjoint

The original dimension was chosen as $n = 1405$. The Figures 9.5 - 9.7 show the decrease of the singular values corresponding to the state equation, the adjoint equation and the velocity-dependent matrices. Each red line in the figures marks where the singular values have been cutted. The cut is based on heuristics, for a given tolerance ϵ_{tol}, we choose the index of

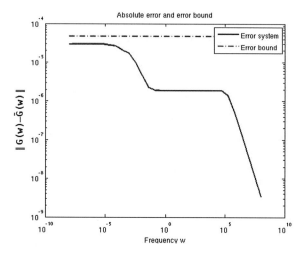

Figure 9.1: Error in the transfer function.

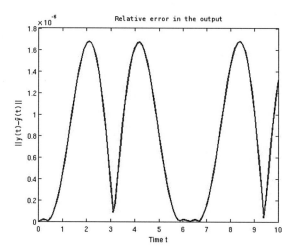

Figure 9.2: Relative error in the output.

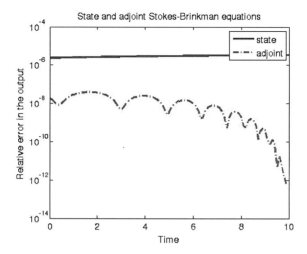

Figure 9.3: Relative error in the output.

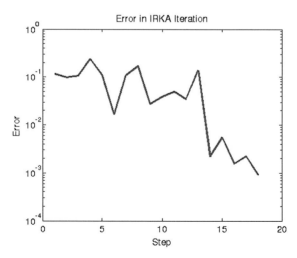

Figure 9.4: Error in the IRKA iteration process.

reduction ℓ as first index where $\frac{\sigma_\ell}{\sigma_1} \leq \epsilon_{\text{tol}}$ is fulfilled. Due to the steep decline of the singular values, the good performance of Matrix DEIM is no surprise. The reduced dimensions were $\ell_{\text{state}} = \ell_{\text{adj}} = 6$ for both the state and the adjoint equation. Due to Matrix DEIM, from all the nonzero matrix entries, only 6 have to be computed for $A(\mathbf{v})$. Figure 9.8 and 9.9 show the relative error between the original equation and the POD-reduced equation as well as the error between the original dimension and the POD and Matrix DEIM reduced equation for state and adjoint advection-diffusion equations, respectively. The difference between both lines can hardly be seen, hence Matrix DEIM works really well for this approach. The following tables show the times for model reduction, assembling of the matrices and time for solving the equations in seconds. The difference in computational time between solving the POD-reduced and POD + MDEIM reduced equations is quite small, but the computational times where measured without the assembling of $A(\mathbf{v})$. The true advantage of using Matrix DEIM occurs in the assembly of the matrix $A(\mathbf{v})$ which can be seen in the second table.

Time for solving in seconds	State	Adjoint
Original	0.5889	0.6068
Reduced with POD	0.0016	0.0067
Reduced with POD + Matrix DEIM	0.0014	0.0061

	Assembling without Matrix DEIM	Matrix DEIM
Time in seconds	221.0869	16.9106

	POD	Matrix DEIM
Time for computing the reduced order models in seconds	0.1218	13.8321

9.2 The optimal control problem

We compared the original optimal control problem to the reduced one. Figure 9.10 shows the difference in the value between the original and the reduced functional, in Figure 9.11 the absolute error is shown. Finally, Figure 9.12 shows the 2−norm of the optimal control.

The following table shows a comparison of the computational costs.

	Original	Reduced
Time for the gradient method in hours	2.7923	0.2084

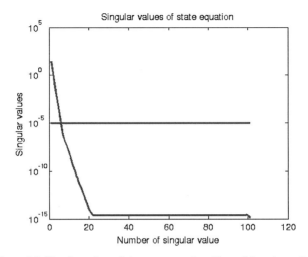

Figure 9.5: Singular values of the state equation. The red line shows the cut.

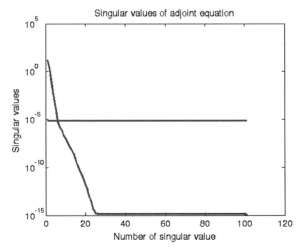

Figure 9.6: Singular values of the adjoint equation. The red line shows the cut.

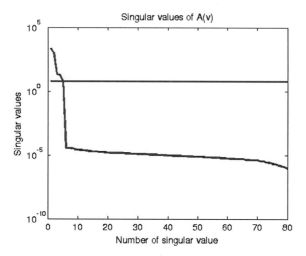

Figure 9.7: Singular values of $A(\mathbf{v})$. The red line shows the cut.

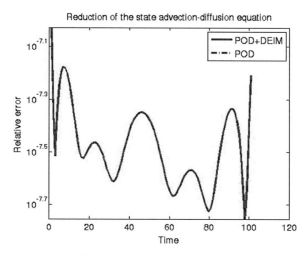

Figure 9.8: Relative error of the state.

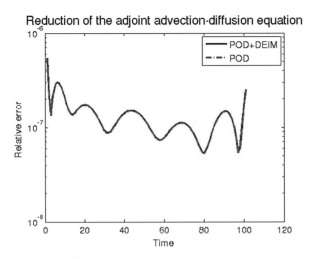

Figure 9.9: Relative error of the adjoint.

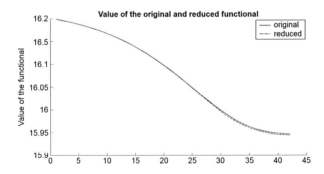

Figure 9.10: Value of the original and reduced functional in each step of the gradient method.

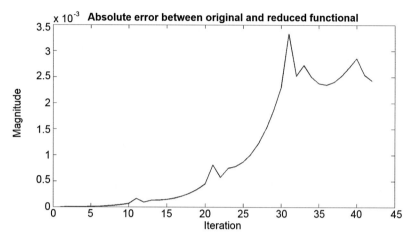

Figure 9.11: Absolute error between the original and reduced functional in each step of the gradient method.

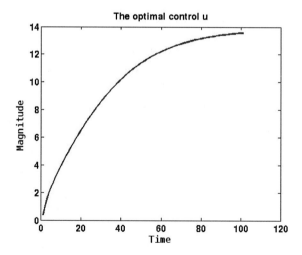

Figure 9.12: The 2-norm of the optimal control **u** on the time interval.

Bibliography

[1] H. Antil, M. Heinkenschloss, and R.H.W. Hoppe. "Domain decomposition and balanced truncation model reduction for shape optimization of the Stokes system". In: *Optim. Methods Softw.* 26.4-5 (2011), pp. 643–669.

[2] H. Antil et al. "Domain Decomposition and Model Reduction for the numerical solution of PDE constrained optimization problems with localized optimization variables". In: *Comput. Vis. Sci.* 13.6 (2010), pp. 249–264.

[3] A.C. Antoulas. *Approximation of Large-scale Dynamical Systems*. Advances in Design and Control. SIAM, 2005.

[4] M. Avriel. *Nonlinear Programming: Analysis and Methods*. Dover Books on Computer Science Series. Dover Publications, 2003.

[5] S. Banach. "Sur les opérations dans les ensembles abstraits et leur application aux équations intégrales". In: *Fund. Math.* 3.1 (1922), pp. 133–181.

[6] C.E. Baukal, V. Gershtein, and X.J. Li. *Computational Fluid Dynamics in Industrial Combustion*. Industrial Combustion. Taylor & Francis, 2000.

[7] R. Becker and B. Vexler. "Optimal control of the convection-diffusion equation using stabilized finite element methods". In: *Numer. Math.* 106.3 (2007), pp. 349–367.

[8] P. Benner, P. Kürschner, and J. Saak. "A reformulated low-rank ADI iteration with explicit residual factors". In: *PAMM* 13.1 (2013), pp. 585–586.

[9] P. Benner, J. Saak, and M.M. Uddin. *Balancing based model reduction for structured index-2 unstable descriptor systems with application to flow control*. Tech. rep. MPIMD/14-20. Max Planck Institute Magdeburg, 2014.

[10] P. Benner and A. Schneider. *Balanced truncation for descriptor systems with many terminals*. Max Planck Institute Magdeburg Preprint MPIMD/13-17. 2013.

[11] P. Benner and A. Schneider. *Balanced Truncation for Descriptor Systems with Many Terminals*. Tech. rep. MPIMD/13-17. Max Planck Institute Magdeburg, 2013.

[12] P. Benner et al. "Low-rank solvers for unsteady Stokes–Brinkman optimal control problem with random data". In: MPIMD/15-10 (2015).

[13] G. Berkooz. "Observations on the proper orthogonal decomposition". In: *Studies in Turbulence*. Springer, 1992, pp. 229–247.

[14] J. Betts and S. Campbell. "Discretize then optimize". In: *Mathematics for Industry: Challenges and Frontiers*. Vol. 1. SIAM, 2005, pp. 27–34.

[15]   H. Brinkman. "A calculation of the viscous force exerted by a flowing fluid on a dense swarm of particles". In: *Appl. Sci. Res.* 1 (1949), pp. 27–34.

[16]   A. N. Brooks and T. J. R. Hughes. "Streamline Upwind/Petrov-Galerkin formulations for convection dominated flows with particular emphasis on the incompressible Navier–Stokes equations". In: *Comput. Methods Appl. Mech. Engrg.* 32 (1990), pp. 199–259.

[17]   S. Chaturantabut and D.C. Sorensen. "Nonlinear model reduction via discrete empirical interpolation". In: *SIAM J. Sci. Comput.* 32.5 (2010), pp. 2737–2764.

[18]   Earl A. Coddington and Norman Levinson. *Theory of Ordinary Differential Equations.* Robert E. Krieger, 1984.

[19]   S.S. Collis and M. Heinkenschloss. *Analysis of the Streamline Upwind/Petrov Galerkin method applied to the solution of optimal control problems.* Tech. rep. TR02–01. Department of Computational and Applied Mathematics, 2002.

[20]   J.B. Conway. *A course in functional analysis.* Graduate texts in mathematics. Springer, 1985.

[21]   B. Dacorogna. *Direct Methods in the Calculus of Variations.* Applied Mathematical Sciences. Springer, 2007.

[22]   T. Damm. "Direct methods and ADI-preconditioned Krylov subspace methods for generalized Lyapunov equations". In: *Numer. Linear Algebra Appl.* 15.9 (2008), pp. 853–871.

[23]   M.A. Day. "The no-slip condition of fluid dynamics". In: *Erkenntnis* 33.3 (1990), pp. 285–296.

[24]   J. E. Dennis Jr. and R. B. Schnabel. *Numerical Methods for Unconstrained Optimization and Nonlinear Equations.* SIAM, 1996.

[25]   *Ecplipse-Af4-Field-Flow-Fractionation-System.* 2016. URL: http://www.wyatt.com/products/instruments/eclipse-af4-field-flow-fractionation-system.html.

[26]   D.F. Enns. "Model reduction with balanced realizations: An error bound and a frequency weighted generalization". In: *Proceedings of the 23rd IEEE Conference on Decision and Control.* 1984, pp. 127–132.

[27]   L.C. Evans. *Partial differential equations.* Graduate studies in mathematics. American Mathematical Society, 1998.

[28]   G. Flagg, C. Beattie, and S. Gugercin. "Convergence of the iterative rational Krylov algorithm". In: *Syst. Control Lett.* 61.6 (2011), pp. 688–691.

[29]   G. B. Folland. *Real analysis.* Pure and Applied Mathematics. John Wiley & Sons, 1999.

[30]   I.M. Gelfand and G.E. Shilov. *Generalized functions.* Academic Press, 1968.

[31]   T. Gelhard et al. "Stabilized finite element schemes with LBB-stable elements for incompressible flows". In: *J. Comput. Appl. Math.* 177.2 (2005), pp. 243–267.

[32] T. H. Gronwall. "Note on the derivatives with respect to a parameter of the solutions of a system of differential equations". In: *Ann. Math.* 20.2 (1919), pp. 292–296.

[33] C. Großmann and H.G. Roos. *Numerische Behandlung Partieller Differentialgleichungen.* Teubner Studienbücher Mathematik. Teubner, 2005.

[34] S. Gugercin. "An iterative rational Krylov algorithm for optimal $\mathcal{H}_2$ model reduction". In: *Proc. Householder Symposium XVI.* 2005.

[35] S. Gugercin and A.C. Antoulas. "A survey of model reduction by balanced truncation and some new results". In: *Internat. J. Control* 77.8 (2004), pp. 749–766.

[36] S. Gugercin, A.C. Antoulas, and C. Beattie. "A rational Krylov iteration for optimal $\mathcal{H}_2$ model reduction". In: *Proceedings of 17th International Symposium on Mathematical Theory of Networks and Systems.* 2006.

[37] S. Gugercin, A.C. Antoulas, and C. Beattie. "Interpolatory model reduction of large-scale dynamical systems". In: *Efficient Modeling and Control of Large-Scale Systems.* Springer, 2010, pp. 3–58.

[38] S. Gugercin, A.C. Antoulas, and C. Beattie. "$\mathcal{H}_2$ model reduction for large-scale linear dynamical systems". In: *SIAM. J. Matrix Anal. Appl.* 30 (2008), pp. 609–638.

[39] S. Gugercin, T. Stykel, and S. Wyatt. "Model reduction of descriptor systems by interpolatory projection methods". In: *SIAM J. Sci. Comput.* 35.5 (2013), B1010–B1033.

[40] J. Guzmán, A.J. Salgado, and F.J. Sayas. "A note on the Ladyženskaja–Babuška–Brezzi condition". In: *J. Sci. Comput.* 56.2 (2013), pp. 219–229.

[41] G. Hadley. *Nonlinear and dynamic programming.* Addison-Wesley Pub. Co., 1964.

[42] Y. Halevi. "Frequency weighted model reduction via optimal projection". In: *IEEE Trans. Autom. Control* 37.10 (1992), pp. 1537–1542.

[43] M. Heinkenschloss, D.C. Sorensen, and K. Sun. "Balanced truncation model reduction for a class of descriptor systems with application to the Oseen equations". In: *SIAM J. Sci. Comput.* 30.2 (2008), pp. 1038–1063.

[44] M. Hinze and S. Volkwein. "Error estimates for abstract linear-quadratic optimal control problems using proper orthogonal decomposition". In: *Comput. Optim. Appl.* 39.3 (2008), pp. 319–345.

[45] D.C. Hyland and D. Bernstein. "The optimal projection equations for model reduction and the relationships among the methods of Wilson, Skelton, and Moore". In: *IEEE Trans. Autom. Control* 30.12 (1985), pp. 1201–1211.

[46] I. Jaimoukha and E.M. Kasenally. "Krylov subspace methods for solving large Lyapunov equations". In: *SIAM J. Numer. Anal.* 31.1 (1994), pp. 227–251.

[47] I.T. Jolliffe. *Principal Component Analysis.* Springer, 1986.

[48] E. Kammann, F. Tröltzsch, and S. Volkwein. "A posteriori error estimation for semilinear parabolic optimal control problems with application to model reduction by POD". In: *ESAIM-Math. Model Num.* 47 (02 2013), pp. 555–581.

[49] K. Karhunen. "Über lineare Methoden in der Wahrscheinlichkeitsrechnung". In: *Annales. Mathematica. Acad. Sci. Fennica.* A1 Mathematics and Physics 37 (1946), pp. 3–79.

[50] G. Kerschen et al. "The method of proper orthogonal decomposition for dynamical characterization and order reduction of mechanical systems: An overview". In: *Nonlinear Dynam.* 41 (2005), pp. 147–169.

[51] D. Kosambi. "Statistics in function space". In: *J. Ind. Math. Soc.* 7 (1943), pp. 76–88.

[52] M. Loeve. "Fonctions Al' eatoires du Second Ordre". In: *Processus stochastiques et mouvement Brownien.* Gauthier-Villars, 1948.

[53] A. Lu and E.L. Wachspress. "Solution of Lyapunov equations by alternating direction implicit iteration". In: *Comput. Math. Appl.* 21.9 (1991), pp. 43–58.

[54] L. Meier and D.G. Luenberger. "Approximation of linear constant systems". In: *IEEE Trans. Autom. Control* 12.5 (1967), pp. 585–588.

[55] B. Moore. "Principal component analysis in linear systems: Controllability, observability, and model reduction". In: *IEEE Trans. Autom. Control* 26.1 (1981), pp. 17–32.

[56] K.W. Morton. *Numerical Solution of Convection-Diffusion Problems.* Appl. Math. Comput. Chapman and Hall, 1996.

[57] C.T. Mullis and R.A. Roberts. "Synthesis of minimum roundoff noise fixed point digital filters". In: *IEEE Trans. Circuits Syst. I. Regul. Pap.* 23.9 (1976), pp. 551–562.

[58] M.A. Obukhov. "Statistical description of continuous fields". In: *Inst, Akad. Nauk SSSR* 24.3 (1954), pp. 3–42.

[59] L. Pernebo and L. Silverman. "Model reduction via balanced state space representations". In: *IEEE Trans. Autom. Control* 27.2 (1982), pp. 382–387.

[60] V.S. Pugachev. "General theory of the correlations of random functions". In: *Izv. Akad. Nauk SSSR Ser. Mat.* 17.5 (1953), pp. 1401–1402.

[61] R. Rannacher. "Methods for numerical flow simulation". In: *Hemodynamical Flows.* Vol. 37. Oberwolfach Seminars. Birkhäuser Basel, 2008, pp. 275–332.

[62] W. Rudin. *Functional analysis.* International Series in Pure and Applied Mathematics. McGraw-Hill Inc., 1991.

[63] W. Rudin. *Real and complex analysis.* McGraw-Hill Book Co., 1987.

[64] H.L. Shang. *A survey of functional principal component analysis.* Tech. rep. 2011.

[65] E.D. Sontag. *Mathematical Control Theory: Deterministic Finite Dimensional Systems.* 2nd ed. Springer, 1998.

[66] T. Stykel. "Balanced truncation model reduction for semidiscretized Stokes equation". In: *Linear Algebra Appl.* 415.2-3 (2006), pp. 262–289.

[67] T. Stykel and V. Simoncini. "Krylov subspace methods for projected Lyapunov equations". In: *Appl. Numer. Math.* 62.1 (2012), pp. 35–50.

[68]   A. Vernhet et al. "Characterization of oxidized tannins: comparison of depolymerization methods, asymmetric flow field-flow fractionation and small-angle X-ray scattering". In: *Anal. Bionanal. Chem.* 401.5 (2011), pp. 1559–1569.

[69]   S. Volkwein. "Model reduction using proper orthogonal decomposition". In: *Lecture Notes Institute of Mathematics and Scientific Computing University of Graz* 7.2 (2008), pp. 1–42.

[70]   S. Volkwein. *Proper orthogonal decomposition and singular value decomposition*. Tech. rep. Graz University, 1999.

[71]   E.F. Whittlesey. "Analytic functions in Banach spaces". In: *Proc. Amer. Math. Soc.* 16.5 (1965), pp. 1077–1083.

[72]   D.A. Wilson. "Optimum solution of model reduction problem". In: *Proc. Inst. Elec. Eng.* 117.6 (1970), pp. 1161–1165.

[73]   W.-Y. Yan and J. Lam. "An approximate approach to $\mathcal{H}_2$ optimal model reduction". In: *IEEE Trans. Autom. Control* 44.7 (1999), pp. 1341–1358.

[74]   E. Zeidler. *Applied Functional Analysis: Main Principles and Their Applications*. Applied Mathematical Sciences. Springer, 1995.